全国高等中医药院校中药学类专业双语规划教材

Bilingual Planned Textbooks for Chinese Materia Medica Majors in TCM Colleges and Universities

分析化学

Analytical Chemistry

（供中药学类、药学类及相关专业使用）

(For Chinese Materia Medica, Pharmacy and other related majors)

主　　编　张　梅

副 主 编　张　娟　詹雪艳　巩丽虹　高晓燕

编　　者（以姓氏笔画为序）

巩丽虹（牡丹江医学院）

朱晓静（山东中医药大学）

孙　飞（广东药科大学）

张　娟（河南中医药大学）

张　梅（成都中医药大学）

陈　慧（湖南中医药大学）

陈美玲（天津中医药大学）

袁　欣（成都中医药大学）

贾明艳（成都中医药大学）

高晓燕（北京中医药大学）

程芳芳（南京中医药大学）

詹雪艳（北京中医药大学）

学术秘书　袁　欣

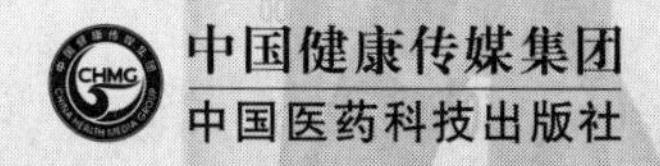

内 容 提 要

本教材系“全国高等中医药院校中药学类专业双语规划教材”之一。全书分为九章，包括绪论、分析误差和数据处理、重量分析法、滴定分析法、酸碱滴定法、沉淀滴定法、配位滴定法、氧化还原滴定法、电位法及双指示电极电流滴定法。每章首尾处分别设置了“学习目标”和“重点小结”，章末附有思考题与习题，书末附有参考答案，方便复习与练习。本教材为书网融合教材，配备有 PPT、习题等数字资源。

本教材内容简明扼要，重点突出，理论联系实际，符合课程要求。可作为高等医药院校中药学、药学、制药技术、制药工程类等专业双语教学及留学生教学使用，也适合化学、食品等其他相关专业学生双语教学使用，同时可供相关科研单位或药品食品质量检验部门科研技术人员参阅。

图书在版编目（CIP）数据

分析化学：汉英对照 / 张梅主编 . —北京：中国医药科技出版社，2020.9

全国高等中医药院校中药学类专业双语规划教材

ISBN 978-7-5214-1883-5

Ⅰ. ①分… Ⅱ. ①张… Ⅲ. ①分析化学 – 双语教学 – 中医学院 – 教材 – 汉、英 Ⅳ. ①O65

中国版本图书馆 CIP 数据核字（2020）第 101880 号

美术编辑 陈君杞

版式设计 辰轩文化

出版 **中国健康传媒集团** | 中国医药科技出版社

地址 北京市海淀区文慧园北路甲 22 号

邮编 100082

电话 发行：010-62227427 邮购：010-62236938

网址 www.cmstp.com

规格 889 × 1194 mm 1/16

印张 14

字数 337 千字

版次 2020 年 9 月第 1 版

印次 2020 年 9 月第 1 次印刷

印刷 三河市万龙印装有限公司

经销 全国各地新华书店

书号 ISBN 978-7-5214-1883-5

定价 49.00 元

出版说明

近些年随着世界范围的中医药热潮的涌动，来中国学习中医药学的留学生逐年增多，走出国门的中医药学人才也在增加。为了适应中医药国际交流与合作的需要，加快中医药国际化进程，提高来中国留学生和国际班学生的教学质量，满足双语教学的需要和中医药对外交流需求，培养优秀的国际化中医药人才，进一步推动中医药国际化进程，根据教育部、国家中医药管理局、国家药品监督管理局等部门的有关精神，在本套教材建设指导委员会主任委员成都中医药大学彭成教授等专家的指导和顶层设计下，中国医药科技出版社组织全国50余所高等中医药院校及附属医疗机构约420名专家、教师精心编撰了全国高等中医药院校中药学类专业双语规划教材，该套教材即将付梓出版。

本套教材共计23门，主要供全国高等中医药院校中药学类专业教学使用。本套教材定位清晰、特色鲜明，主要体现在以下方面。

一、立足双语教学实际，培养复合应用型人才

本套教材以高校双语教学课程建设要求为依据，以满足国内医药院校开展留学生教学和双语教学的需求为目标，突出中医药文化特色鲜明、中医药专业术语规范的特点，注重培养中医药技能、反映中医药传承和现代研究成果，旨在优化教育质量，培养优秀的国际化中医药人才，推进中医药对外交流。

本套教材建设围绕目前中医药院校本科教育教学改革方向对教材体系进行科学规划、合理设计，坚持以培养创新型和复合型人才为宗旨，以社会需求为导向，以培养适应中药开发、利用、管理、服务等各个领域需求的高素质应用型人才为目标的教材建设思路与原则。

二、遵循教材编写规律，整体优化，紧跟学科发展步伐

本套教材的编写遵循“三基、五性、三特定”的教材编写规律；以“必需、够用”为度；坚持与时俱进，注意吸收新技术和新方法，适当拓展知识面，为学生后续发展奠定必要的基础。实验教材密切结合主干教材内容，体现理实一体，注重培养学生实践技能训练的同时，按照教育部相关精神，增加设计性实验部分，以现实问题作为驱动力来培养学生自主获取和应用新知识的能力，从而培养学生独立思考能力、实验设计能力、实践操作能力和可持续发展能力，满足培养应用型和复合型人才的要求。强调全套教材内容的整体优化，并注重不同教材内容的联系与衔接，避免遗漏和不必要的交叉重复。

三、对接职业资格考试，“教考”“理实”密切融合

本套教材的内容和结构设计紧密对接国家执业中药师职业资格考试大纲要求，实现教学与考试、理论与实践的密切融合，并且在教材编写过程中，吸收具有丰富实践经验的企业人员参与教材的编写，确保教材的内容密切结合应用，更加体现高等教育的实践性和开放性，为学生参加考试和实践工作打下坚实基础。

四、创新教材呈现形式，书网融合，使教与学更便捷更轻松

全套教材为书网融合教材，即纸质教材与数字教材、配套教学资源、题库系统、数字化教学服务有机融合。通过“一书一码”的强关联，为读者提供全免费增值服务。按教材封底的提示激活教材后，读者可通过PC、手机阅读电子教材和配套课程资源（PPT、微课、视频等），并可在线进行同步练习，实时收到答案反馈和解析。同时，读者也可以直接扫描书中二维码，阅读与教材内容关联的课程资源，从而丰富学习体验，使学习更便捷。教师可通过PC在线创建课程，与学生互动，开展在线课程内容定制、布置和批改作业、在线组织考试、讨论与答疑等教学活动，学生通过PC、手机均可实现在线作业、在线考试，提升学习效率，使教与学更轻松。此外，平台尚有数据分析、教学诊断等功能，可为教学研究与管理提供技术和数据支撑。需要特殊说明的是，有些专业基础课程，例如《药理学》等9种教材，起源于西方医学，因篇幅所限，在本次双语教材建设中纸质教材以英语为主，仅将专业词汇对照了中文翻译，同时在中国医药科技出版社数字平台“医药大学堂”上配套了中文电子教材供学生学习参考。

编写出版本套高质量教材，得到了全国知名专家的精心指导和各有关院校领导与编者的大力支持，在此一并表示衷心感谢。希望广大师生在教学中积极使用本套教材和提出宝贵意见，以便修订完善，共同打造精品教材，为促进我国高等中医药院校中药学类专业教育教学改革和人才培养做出积极贡献。

全国高等中医药院校中药学类专业双语规划教材建设指导委员会

数字化教材编委会

主　　编　张　梅

副 主 编　张　娟　詹雪艳　巩丽虹　高晓燕

编　　者　（以姓氏笔画为序）

巩丽虹（牡丹江医学院）

孙　飞（广东药科大学）

张　梅（成都中医药大学）

陈美玲（天津中医药大学）

贾明艳（成都中医药大学）

程芳芳（南京中医药大学）

朱晓静（山东中医药大学）

张　娟（河南中医药大学）

陈　慧（湖南中医药大学）

袁　欣（成都中医药大学）

高晓燕（北京中医药大学）

詹雪艳（北京中医药大学）

学术秘书　袁　欣

前 言

在经济全球化、教育国际化的大背景下，依据教育部关于高等学校本科教学质量工程要重视双语教学的文件精神，为培养具有国际视野和国际交流合作意识的国际化中医药人才，提高来华中药学类专业留学生的教育质量，组织开展了《分析化学》双语教材的编写。

本教材以英文原版教材为主、中文教材为参考进行编写，尽量保持英文版教材的“原汁原味”，亦适应国内教学实际需求，使学生在学习掌握分析化学基本理论和方法的同时提高英文的阅读理解水平。首先按照中文版教材特有的系统性和条理性对英文原版教材的章节顺序进行整合调整，以符合国内学生的阅读习惯；其次对英文原版教材内容进行精炼，保留其编写生动详细、语句通俗易懂的基础性重点章节内容，而对非重点内容、插图和例题适当删减；同时为方便学生对课程内容的理解、记忆及自学，对双语教材中出现的分析化学专用词汇、术语添加了中文注释。在每章开头设置了“学习目标”，从知识和能力两方面提出每章学习要求，章末设有中文的“重点小结”，帮助学生梳理该章节主要内容，掌握重点和难点。本教材是“书网融合”教材，该板块主要由电子教材（含中文电子教材）、教学配套资源（PPT、微课、视频、图片等）、题库系统、数字化教学系统（在线教学、在线作业、在线考试）等组成，帮助学生全面掌握教材内容，评价学习成效。

本教材由全国9所中医药院校长期从事分析化学教学工作，并具有良好英文写作水平的一线教师编写，各编委分章节独立编写，经主编、副主编初审及修改，由主编及学术秘书整理定稿。

本教材可供全国高等中医药院校中药学、药学专业双语教学及留学生教学使用，也适合化学、食品等其他相关专业双语教学使用，同时可供相关专业双语教学和科研人员参阅。

本书编写分工如下：第一章张梅，第二章袁欣、贾明艳，第三章陈美玲、孙飞，第四章朱晓静，第五章詹雪艳、高晓燕，第六章陈慧，第七章巩丽虹，第八章张娟，第九章程芳芳。

在本书编写过程中得到各编委所在院校大力支持，在此表示最诚挚的谢意。同时，在编写过程中，编者参阅了相关书籍和资料，在此向作者表示深深的谢意。

限于编者水平和经验，书中可能存在疏漏和不足，恳请有关专家、同行和同学提出宝贵意见，以便再版修订。

编 者

2020年4月

Preface

Based on the gist of document from the Ministry of Education which emphasized the importance of bilingual education in undergraduate teaching quality program of universities, the bilingual textbook *Analytical Chemistry* is compiled and organized by China Medical Science Press under the background of economic globalization and globalization of education in order to cultivate internationalized Chinese medicine talents with international view as well as awareness of international communication and cooperation, and improve the quality of education for overseas students whose majors are related to Chinese medicine.

Analytical Chemistry is mainly based on original English textbook and also takes Chinese textbook for reference. We keep it as authentic as possible and also adapt it to meet the actual teaching needs in China, in this way, the students can improve their English reading skills while learning and understanding the basic theories and methods of analytical chemistry. Firstly, we adjust the order of the chapters and integrate some contents in the original English textbook based on the systemic and organized characteristics of Chinese textbook to make it in accord with Chinese students' reading habit. Secondly, we keep the basic and vital chapters which are vivid, specific and easy to understand while deleting the less important content, illustration and examples appropriately in order to refine the content in the original English textbook; at the same time, we add Chinese annotations to analytical chemistry words and terms appeared in the bilingual textbook to make it convenient for students to understand, memorize and study the course content independently. There are "learning objectives" in the beginning of every chapter to raise learning requirements from the aspects of both knowledge and ability and also "key points summary" in Chinese in the end of each chapter to help students clear out the main content of the chapter and to grasp the key and difficult points. This textbook is also supported with "book & internet convergence" module, which is mainly consisted of the PowerPoint, learning tips and workbook (including single choice, multiple choice, true/false, gap-fill exercise and so on) for every chapter and is closely corresponded to the textbook. This module is used to help students to absorb textbook content and assess learning performance comprehensively.

This textbook is compiled by front-line teachers from nine Chinese medicine universities who have long engaged in analytical chemistry with great English writing skills. Every chapter is compiled by respective editorial board member independently, checked and modified by editor-in-chief and associate editor, then finalized by editor-in-chief and editorial secretary.

This textbook can be used in bilingual analytical chemistry teaching for undergraduates and overseas students in majors like Traditional Chinese medicine, Pharmacy as well as other related majors like Chemistry and Food. It can also be a reference for teachers and researchers in related majors.

The division of labor of the compiling is as follows: Chapter 1 by Zhang Mei, Chapter 2 by Yuan Xin, Jia Mingyan, Chapter 3 by Chen Meiling, Sunfei, Chapter 4 by Zhu Xiaojing, Chapter 5 by Zhan Xueyan, Gao Xiaoyan, Chapter 6 by Chen Hui, Chapter 7 by Gong Lihong, Chapter 8 by Zhang Juan and Chapter 9 by Cheng Fangfang.

In the process of compiling this textbook, we would like to express our sincerest acknowledgement for the great support offered by universities where all editorial boards came from. Simultaneously, in the process of compiling, our editors referred to related books and materials, we would like to show our deepest gratitude here.

Due to the limited standard and experience of editors, there might be mistakes and deficiencies in this textbook, we sincerely welcome criticism and correction from related professors, peers and students.

Editors

April 2020

Contents

Chapter 1 Introduction ······ 1

Section 1 The Nature and Role of Analytical Chemistry ······ 1
Section 2 The Classification of Analytical Chemistry ······ 2
Section 3 The Total Analytical Process ······ 4
Section 4 The History of Analytical Chemistry ······ 6
Section 5 The Literature of Analytical Chemistry ······ 8

Chapter 2 Errors and Data Treatment in Quantitative Analysis ······ 11

Section 1 Errors in Quantitative Analysis ······ 11
Section 2 Significant Figures ······ 16
Section 3 Statistical Treatment of Analytical Data ······ 18
Section 4 Methods to Improve the Accuracy of Analysis Results ······ 28

Chapter 3 Gravimetric Analysis ······ 33

Section 1 Introduction to Gravimetric Analysis ······ 33
Section 2 Volatilization Gravimetry ······ 34
Section 3 Extraction Gravimetry ······ 35
Section 4 Precipitation Gravimetry ······ 36

Chapter 4 Titrimetric Analysis ······ 49

Section 1 Introduction to Titrimetric Analysis ······ 49
Section 2 Primary Standards and Standard Solutions ······ 52
Section 3 The Calculations in Titrimetric Analysis ······ 55

Chapter 5 Acid-Base Titrations ······ 60

Section 1 Introduction of Acid-Base Titrations ······ 60
Section 2 Acid-Base Equilibria in Aqueous Solutions ······ 61
Section 3 Acid-Base Indicators ······ 71
Section 4 Acid-Base Titration Curve and Indicator Selection ······ 74
Section 5 Applications of Acid-Base Titrations ······ 87
Section 6 Acid-Base Titrations in Nonaqueous Solutions ······ 93

Chapter 6 Precipitation Titration ······ 107

Section 1 Introduction to Precipitation Titration ······ 107
Section 2 Argentometry ······ 107

Section 3 Application Examples …… 114

Chapter 7 Complexometric Titration …… 119

Section 1 Basic Principles of Complexometric Titration …… 119
Section 2 Improving the Selectivity of Complexometric Titration …… 135
Section 3 Complexometric Titration Method and Its Application …… 136

Chapter 8 Redox Titrations …… 143

Section 1 Redox Reactions …… 143
Section 2 Redox Titration Curves …… 148
Section 3 Detection of End Point: Indicators …… 151
Section 4 Methods Involving Iodine: Iodimetry and Iodometry …… 153
Section 5 Redox Titrations with Other Oxidants …… 156
Section 6 Quantitative Calculations for Redox Titrations …… 160

Chapter 9 Potentiometry and Biamperometric Titration Method …… 164

Section 1 Introduction to Electrochemical Analysis …… 164
Section 2 Basic Principles of Potentiometry …… 165
Section 3 Reference Electrode and Indicator Electrode …… 167
Section 4 Direct Potentiometry …… 172
Section 5 Potentiometric Titration …… 180
Section 6 Biamperometric Titration …… 183

Answer to Exercises …… 188

Appendix …… 192

Appendix 1 Atomic Weight …… 192
Appendix 2 Formula Weights …… 194
Appendix 3 Dissociation Constants for Acids (25℃) …… 196
Appendix 4 Dissociation Constants for Bases (25℃) …… 199
Appendix 5 Standard Electrode Potential (18–25℃) …… 200
Appendix 6 Conditional Electrode Potential …… 204
Appendix 7 Solubility Product Constants (18–25℃, I=0) …… 206

参考文献 …… 209

Chapter 1 Introduction

学习目标

知识要求

1. **掌握** 分析化学的性质及分类方法。
2. **熟悉** 定量分析的一般步骤。
3. **了解** 分析化学发展进程。

能力要求

通过学习分析化学的性质及相关研究内容，为后续各章节顺利展开作铺垫，亦为药物分析等专业课程的学习奠定基础。

Section 1 The Nature and Role of Analytical Chemistry

Analytical chemistry is a scientific discipline that develops and applies methods, instruments, and strategies to obtain information on the composition and nature of matter in space and time (Edinburgh-Definition of WPAC 1993).

It is an important branch of chemistry, is a science that investigates the analytical methods and relevant theories of composition, content, structure, species and other chemical information of substance. The science, taking chemical fundamental theories and laboratory technologies as the basis, extensively absorbs and fuses knowledge from physics, biology, mathematics, computer science, statistics and informatics. It develops to provide substance information data source necessary for science and technology advancement.

Of the analytical chemistry, the prime task is to acquire images, data and other relevant information by different methods and approaches; to identify the chemical composition of a material system; to measure the content of key ingredients, and also to determine the material structure and species of the system. Mainly, it covers qualitative analysis, quantitative analysis, structural analysis and species analysis.

Analytical chemistry is used in many fields. It not only plays a critical part in the development of the discipline itself but also holds a decisive position in various respects, including national economy, science and technology, medicine and health, as well as school education.

Throughout the progress of chemistry, analytical chemistry contributes a lot of remarkable achievements. Those contributions infuse everywhere, inextricable to the establishment of element discovery and different chemical basic laws (law of conservation of mass, law of constant proportions, etc.); the determination of relative atomic masses; the foundation of periodic law of elements, the discovery of element characteristic spectral lines, and other disclosures of chemical phenomena. "Since the presence of science and technology among mankind, there has been chemistry and analytical chemistry is its start point."

In the construction of national economy, analytical chemistry is irreplaceable. It functions to provide data and information relevant to resource exploration, reserve determination of natural gas, oil and mineral resources; to location selection of coal mine and iron and steel bases; to selection of raw materials for industrial production, and inspection of intermediates, finished products and related substances; to verification of soil composition in agricultural production, diagnosis of crop nutrition, and quality testing of agricultural products and processed food; to assessment on the character, mechanical strength and quality of various types of architectures and decorative materials in the construction industry; also, to quality monitoring and control of all goods flowing around commercial circulation.

Besides, analytical chemistry has crossed over the whole chemistry in terms of scientific and technological research, reaching to other fields, such as life science, material science, environmental science and energy science, where it exercises a vital role. For example, the quantitative analysis on DNA, protein and carbohydrate and other cell contents may actualize early notice, diagnosis and treatment for diseases like cancers. Minute monitoring and analyses on each human-living environment component, especially on nature, source, content and distribution pattern of some critical hazard pollutants, lays a considerable foundation to understand, evaluate and protect environments. Therefore, analytical chemistry can be hailed as an indispensable study tool and means for any scientific and research fields that involves chemical phenomena. In fact, it has upgraded into "a science engaging in scientific research", functioning as the "eye" of modern science and technology.

Throughout the pharmaceutical industry analytical chemistry leaves its trace everywhere. It is a necessity of clinical disease diagnosis, pathological examination; drug quality control, research of new drugs; isolation and identification of active components in natural drugs; studies on structure-activity and dose-effect relationships of drugs; explorations on in-vivo processes of drugs; researches of drug formulation stability; as well as management of public health emergencies. Analytical chemistry not merely acts to identify problems but also participates in solving practical matters.

Moreover, analytical chemistry is also viewed as a key professional basic discipline of pharmacy-related majors in school education. Its theoretical knowledge and laboratory techniques are universally applied in advanced professional courses (pharmaceutical analysis, pharmaceutical chemistry, pharmacy, pharmacology, natural pharmaceutical chemistry, etc.). Meanwhile, it lays a momentous foundation for students to cultivate their professional competence.

Section 2 The Classification of Analytical Chemistry

According to the analysis task, the analysis object, the determination principle, the difference of the sample dosage and the content of the component to be measured, and the working property, the analytical chemistry method can be classified into various categories.

1. Qualitative analysis, quantitative analysis, structural analysis, and species analysis

According to the classification of analytical tasks, analytical chemistry can be divided into qualitative analysis, quantitative analysis, structural analysis, and species analysis.

Qualitative analysis indicates whether a particular element or compound is in the sample, quantitative analysis gives the amount of the species in the sample, structure analysis study the molecular structure or crystal structure of matter, whereas species analysis refers to the process of determining the atomic and molecular composition forms of analytical substances.

2. Inorganic analysis and organic analysis

According to different analysis objects, analytical chemistry can be divided into the inorganic analysis and organic analysis.

The object of inorganic analysis is the inorganic matter. Because inorganic matter is composed of many elements, it usually requires the identification of the material composition (element, ion, atomic group or compound) and the determination of the content of each component.

The object of organic analysis is organic matter, which is composed of a limited number of elements such as carbon, hydrogen, oxygen, nitrogen, sulfur, and halogen. However, there are millions of species of organic matter in nature and their structure is complex, so the focus of analysis is functional group analysis and structural analysis.

3. Chemical analysis and instrumental analysis

According to the different analysis principles, the analytical chemistry methods can be classified into chemical analysis and instrumental analysis.

Analytical methods based on chemical reactions and their quantitative relationships of substances are called chemical analysis. It is the basis of analytical chemistry, has a long history and is often referred to as the classical analysis method, mainly including gravimetric analysis (metric analysis) and titrimetric analysis (volumetric analysis).

Gravimetric analysis and titrimetric analysis are mainly used for the determination of major components (content is higher than 1%). The former has high accuracy and is still the standard method for the determination of some components; the latter is characterized by simple equipment, simple operation, time-saving, rapid and accurate results (relative error ±0.2%), and it is an important routine analysis method.

Instrumental analysis is an analytical method based on the physical and chemical properties of substances, so it is also called physical analysis and physicochemical analysis.

This kind of method is often carried out by measuring the physical or physical-chemical parameters of a substance, which requires a special instrument, so it is often called instrumental analysis.

4. Macro analysis, semimicro analysis, micro analysis and ultramicro analysis

On the basis of sample size, analytical methods are often classified as macro analysis (the analysis of quantities of 0.1g (10ml) or more); semimicro analysis (dealing with quantities ranging from 0.01g to 0.1g (1ml to 10ml)); micro analysis (for quantities in the range 0.1mg to 10mg (0.01ml to 1ml)); ultramicro analysis (for quantities below 0.1mg (0.01ml)).

5. Routine analysis and arbitral analysis

According to the nature of work, the method of analytical chemistry can be divided into the routine analysis and arbitral analysis. Routine analysis refers to the daily work , such as a pharmaceutical factory following the drug quality standard for each batch of drug production, whereas arbitral analysis is carried out by an authoritative analytical institution when the analytical results are disputed.

Section 3 The Total Analytical Process

Quantitative analysis is one of the main tasks of analytical chemistry. The purpose of quantitative analysis is to determine the content of one or more components in a substance. The jobs usually involved are: ①sampling; ②sample preparation; ③analysis; ④calculating results and evaluation of data; ⑤conclusions and report.

1. Sampling

Sampling is the process of selecting representative material to analyze. To produce meaningful information, an analysis must be performed on a sample that has the same composition as the bulk of material from which it was taken. When the bulk is large and heterogeneous, great effort is required to get a representative sample. Whether sampling is simple or complex, however, the analyst must be sure that the laboratory sample is representative of the whole before proceeding. Sampling is frequently the most difficult step in an analysis and the source of greatest error.

1.1 Solid materials

To make collected samples representative, multiple sampling points should be selected from different parts and depths according to the distribution of sample components and particle sizes. There are a variety of methods for selecting sampling points. For example, the random sampling method, a way to select sampling points randomly, requires more sampling points and thus renders them more representative. The judgment sampling method aims to selectively choose sampling points in accordance with the distribution of components. The systematic sampling method refers to a selection of sampling points according to certain rules. Obviously, a larger number of samples bring higher accuracy but also an increased in cost. Thus, the number of samples should be able to meet the expected requirements while to save cost as much

as possible. For heterogeneous materials, the sample collection depends on the size and specific gravity of particles, sample uniformity and the accuracy of analysis. Larger particles bring higher specific gravity and a bigger minimum collection amount, and a more uneven sample and higher analysis requirements demands greater minimum collection.

1.2 Liquids and gases

Usually, liquids and gases are sufficiently homogeneous or so easily homogenized that can be done with small samples regularly. Small sample vessels hold a large ratio of surface to volume whereupon absorption-induced losses might be large in a relative way. On this account, it is complete rinsing of vessels with liquid or gas for analysis that is a prerequisite for an equilibrium setup between samples and the vessel walls (where the process refers to an "equilibration"). Given the possible influence of the interaction between test samples and sample containers on the practical composition of the latter, container walls are required to be preconditioned with no changes present in the test sample.

2. Sample preparation

Sample preparation is the process of converting a representative sample into a form suitable for chemical analysis, which usually means dissolving the sample. Samples with a low concentration of an analyte may need to be concentrated before analysis. It may be necessary to remove or mask species that interfere with the chemical analysis.

A solid laboratory sample is ground to decrease particle size, mixed to ensure homogeneity, and stored for various lengths of time before analysis begins. Absorption or desorption of water may occur during each step, depending on the humidity of the environment. Because any loss or gain of water changes the chemical composition of solids, it is a good idea to dry samples just before starting an analysis. Alternatively, the moisture content of the sample can be determined at the time of the analysis in a separate analytical procedure.

Liquid samples present a slightly different but related set of problems during the preparation step. If such samples are allowed to stand in open containers, the solvent may evaporate and change the concentration of the analyte. If the analyte is a gas dissolved in a liquid, as in our blood gas example, the sample container must be kept inside a second sealed container, perhaps during the entire analytical procedure, to prevent contamination by atmospheric gases. Extraordinary measures, including sample manipulation and measurement in an inert atmosphere, may be required to preserve the integrity of the sample.

In trace analysis, analyte concentrations are too low in general to determine by any method. Nevertheless, an increase in the fractions of analytes can be realized through enrichment, while their popular techniques are those based on extraction, adsorption, and ion exchange. Normally, separation methods are welcome to the elimination of compounds interfering in determination. Sometimes, the contribution some unwanted components bring to the results of measurement can be inhibited by adding interferent-bound reagents, which is a means of "masking".

3. Analysis

Measure the concentration of analyte in several identical aliquots (portions). The purpose of replicate measurements is to assess the variability (uncertainty) in the analysis and to guard against a gross error in the analysis of a single aliquot. The uncertainty of a measurement is as important as the measurement itself, because it tells us how reliable the measurement is. If necessary, use different analytical methods on similar samples to make sure that all methods give the same result and that the choice of analytical method is not biasing the result. You may also wish to construct and analyze several different bulk samples to see what variations arise from your sampling procedure.

4. Calculating results and evaluation of data

Computing analyte concentrations from experimental data is usually relatively easy, particularly with computers. These computations are based on the raw experimental data collected in the measurement step, the characteristics of the measurement instruments, and the stoichiometry of the analytical reaction.

Analytical results are complete only when their reliability has been estimated. The experimenter must provide some measure of the uncertainties associated with computed results if the data are to have any value.

Most chemical analyses are performed on replicate samples whose masses or volumes have been determined by careful measurements with an analytical balance or with a precise volumetric device. Replication improves the quality of the results and provides a measure of their reliability. Quantitative measurements on replicates are usually averaged, and various statistical tests are performed on the results to establish their reliability.

5. Conclusions and report

Once the concentration of analyte in the prepared sample solution has been determined, the results are used to calculate the amount of analyte in the original sample. Either an absolute or a relative amount may be reported. Usually, a relative composition is given, for example, percent or parts per million, along with the mean value for expressing accuracy. Replicate analyses can be performed (three or more), and a precision of the analysis may be reported, for example, standard deviation. A knowledge of the precision is important because it gives the degree of uncertainty in the result. The analyst should critically evaluate whether the results are reasonable and relate to the analytical problem as originally stated.

Deliver a clearly written, complete report of your results, highlighting any limitations that you attach to them. Your report might be written to be read only by a specialist or it might be written for a general audience. Be sure the report is appropriate for its intended audience.

Section 4 The History of Analytical Chemistry

Analytical chemistry is a long-history science with an origin dated back to ancient alchemy. The development of ancient medicine, agriculture, metal smelting and other technologies demands an understanding of the composition of substances, greatly stimulating the advancement of various qualitative and quantitative detection technologies. Nonetheless, no systematic theory was formed.

It was not until the end of the 19th century when the law of impermanence of matter, periodic law of elements and solution equilibrium theory was founded and boomed that the theoretical foundation of analytical chemistry was laid, evolving analytical chemistry from detection technology into an independent science. In the 20th century, with the rapid development of modern science and technology and the mutual integration and penetration of disciplines, analytical chemistry underwent three significant changes along with the development of chemistry and other related disciplines.

The first step in the evolution of analytical chemistry

Early 20th century witnessed the formation of a theoretical basis for analytical chemistry, a gift from the development of physicochemical solution theory. Furthermore, the establishment of four equilibrium theories (acid-base equilibrium, redox equilibrium, complexation equilibrium, and precipitation equilibrium) in a solution accelerated the ripening and perfection of theories and methods of chemical analysis.

The second step in the evolution of analytical chemistry

From 1940s to 1960s, the development of physics and electronics promoted the establishment of analytical methods based on the physical and physicochemical properties of substances. In the meantime, a variety of simple and rapid instrument analysis methods, such as spectral analysis and polarographic analysis, emerged. With the growth of instrumental analysis methods, classical analytical chemistry based on chemical analysis upgraded to a modern analytical chemistry based on instrumental analysis.

The third step in the evolution of analytical chemistry

Since the late 1970s, the advent of information age, mainly marked by the application of computers, has brought about more profound changes in analytical chemistry. To meet the development needs of life science, environmental science, energy science and material science, the basic theories and test methods of analytical chemistry were improved progressively. Combined with the application of computer technology in image and data processing, it creates conditions for the establishment of new methods with high sensitivity, selectivity and accuracy in analytical chemistry. Breaking through the boundary of pure chemistry, modern analytical chemistry closely combines with mathematics, physics, computer science, and biology, and develops into a comprehensive science characterized by interdisciplinary integration. The requirement for analytical chemistry is no longer limited to the general "what" (qualitative analysis) and "how much" (quantitative analysis), but the capability of providing more comprehensive multidimensional information about substances. With this progression springs up a host of new methods and technologies: particle analysis, micro-area analysis, layer by layer analysis, non-destructive analysis, and dynamic analysis including on-line analysis, real-time analysis, and *in situ* and *in vivo* analysis.

Since the 21st century, science and technology have embraced their blossoming; related improvements and innovations spring up incessantly. Analytical chemistry, while assimilating the latest achievements of science and technology, will continue its advancement in the aspects of high sensitivity (up to molecular and even atomic levels), high selectivity (complex system), high amount of information

(processing of huge or even tremendous amounts of data), accuracy, speed, simplicity and economy; and also develop in both depth and breadth in the areas of miniaturization, automation, digitalization, computerization, intellectualization and informatization. Applications of combined technologies or instruments are reinforced, and a variety of novel analysis methods are established. With these aids, a larger number of newer, more sophisticated subjects will be solved, so as to create greater progress for science and technology development and human progressions.

Section 5 The Literature of Analytical Chemistry

When defining a problem, one of the first things the analyst does is to go to the scientific literature and see if the particular problem has already been solved in a manner that can be employed. There are many reference books in selected areas of analytical chemistry that describe the commonly employed analytical procedures in a particular discipline and also some of the not-so-common ones. These usually give reference to the original chemical journals. For many routine or specific analysis, prescribed standard procedures have been adopted by the various professional societies.

If you do not find a solution to the problem in reference books, then you must resort to the scientific journals. *Chemical Abstracts is* the logical place to begin a literature search. This journal contains abstracts of all papers appearing in the major chemical publications of the world. Yearly and cumulative indexes are available to aid in the literature search. The element or compound to be determined as well as the type of sample to be analyzed can be looked up to obtain a survey of the methods available. Author indexes are also available. School library may subscribe to *SciFinder Scholar*, the online access to *Chemical Abstracts*. You can search by chemical substance, topic, author, company name, and access abstracted journals. Once you have an article online, you can link to referenced articles in the paper.

You can also locate many relevant references for a specific problem by using Web search engines. Following is a selected list of some references in analytical chemistry.

Table 1-1 Monograph

Title	Author	Publisher and time of publication
Quantitative Chemical Analysis (10th ed)	Harris D C	W. H. Freeman and Company, 2019
Analytical Chemistry (7th ed)	Christian G D, Dasgupta P K, Schug K A	Wiley, 2013
Fundamentals of Analytical Chemistry (9th ed)	Skoog D A, West D M, Holler F J, Crouch S R	Brooks/Cole, Cengage Learning, 2013
Analytical Chemistry and Quantitative Analysis	Hage D S, Carr J D	Prentice Hall, 2011
Vogel's textbook of Quantitative Chemical Analysis (6th ed)	Mendham J, Denney R C, Barnes J D, Thomas M J K	Prentice Hall, 2000

continued

Title	Author	Publisher and time of publication
21 世纪的分析化学	汪尔康	北京：科学出版社，1999
定量化学分析(第 2 版)	李龙泉，朱玉瑞，金谷，江万权，邵利民	合肥：中国科学技术大学出版社，2005
分析化学原理(第 2 版)	吴性良，孔继烈	北京：化学工业出版社，2010
化学分析测量不确定度评定应用实例	郝玉林	北京：中国标准出版社，2011

Table 1-2 Journals

Home	Abroad
1. 分析化学	1. Analytical Chemistry (USA)
2. 分析测试学报	2. The Analyst (UK)
3. 分析试验室	3. Analytical Methods (UK)
4. 分析科学学报	4. Analytical Letters (USA)
5. 化学学报	5. Journal of Chromatography (NL)
6. 化学通报	6. Analytical Chimica Acta (NL)
7. 高等学校化学学报	7. Analytical and Bioanalytical Chemistry (GER)
8. 色谱	8. Analytical Sciences (JPN)
9. 光谱学与光谱分析	9. Talanta (UK)
10. 药物分析杂志	10. Trends in Analytical Chemistry (UK)

Table 1-3 Common chemical network database

Website	Publisher
https: //scifinder. cas. org	美国化学文摘服务社
http: //www. interscience. wiley. com	美国 Wiley 公司
http: //www. rsc. org/journals	英国皇家化学学会
http: //pubs. acs. org/	美国化学学会
http: //www. sciencedirect. com	荷兰 Elsevier Science 公司
http: //www. cnki. net	清华大学及清华同方
http: //www. cqvip. com	重庆维普咨询有限公司
http: //www. wanfangdata. com. cn	北京万方数据有限公司

重点小结

分析化学是研究获取物质的组成、含量、结构和形态等化学信息的分析方法及相关理论的一门科学。分析化学的主要任务是通过各种方法与手段，获取图像、数据等相关信息用于鉴定物质体系的化学组成、测定其中有关成分的含量和确定体系中物质的结构与形态。主要内容包括定性分析、定量分析、结构分析和形态分析。分析化学应用广泛，遍及社会发展各个方面，包括国民经济、科学技术、医药卫生、学校教育等方面。

根据分析任务、分析对象、测定原理、试样用量与待测组分含量的不同以及工作性质等，可将分析化学的方法进行多种分类。按分析任务可分为定性分析、定量分析、结构分析、形态分析；按分析对象可分为无机分析和有机分析；按分析原理可分为化学分析和仪器分析；按试样用量可分为常量分析、半微量分析、微量分析、痕量分析；按工作性质可分为常规分析和仲裁分析。

定量分析的基本步骤包括：①样品采集；②样品制备；③样品分析；④分析结果的计算及评价；⑤结论与报告。

分析化学是一门古老的科学，历史悠久，随着化学和其他相关学科的发展而不断发展，经历了三次巨大的变革。第一次变革：20 世纪初，建立溶液四大平衡。第二次变革：20 世纪 40—60 年代，仪器分析出现并快速发展。第三次变革：20 世纪 70 年代至今，分析化学发展成为具有多学科交叉融合特征的综合科学。

题库

目标检测

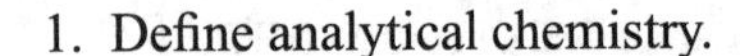

1. Define analytical chemistry.
2. Distinguish between qualitative analysis and quantitative analysis.
3. Describe the classification methods and basis of analytical chemistry.
4. Outline the steps commonly employed in an analytical procedure. Briefly describe each step.

Chapter 2 Errors and Data Treatment in Quantitative Analysis

学习目标

知识要求

1. 掌握 分析误差的种类、产生的原因及其减免方法；准确度与精密度的表示方法及相互关系；有效数字及其运算规则。

2. 熟悉 误差的传递规律，偶然误差的分布；可疑数据的取舍方法；置信区间的定义及表示方法；显著性检验的方法。

3. 了解 相关分析和回归分析。

能力要求

通过学习误差的基础理论和相关知识，对定量分析结果的可靠性和准确度做出合理判断和正确表达。

The task of quantitative analysis is to obtain the accurate content of the determined component in a sample through experiment. In the analytical process, although some experimental errors are more obvious than others, there is error associated with every measurement. It is impossible to measure the "true value" of anything. The best we can do in a chemical analysis is to minimize errors and estimate their size with acceptable accuracy.

Section 1 Errors in Quantitative Analysis

PPT

1. Systematic errors and random errors

The difference between the measured value and the true value is called the error. When the measured value is greater than the real value, the error is positive. Conversely, it is negative.

Errors can be classified into systematic errors and random errors, depending on their characteristics and sources.

1.1 Systematic errors

Systematic errors are caused by some assignable causes, and are constant in both direction and

magnitude in repeated measurement. Theoretically, systematic errors are measurable, so they are often called determinate errors. Some common systematic errors are:

1.1.1 Errors of methods. These arise from imperfect analytical methods. For instance, in a gravimetric analysis, slight solubility of a precipitate will introduce a negative error and coprecipitation of impurities should lead to a positive error. In a titrimetric method, the inconformity between the titration endpoint and stoichiometric point would also result in errors.

1.1.2 Instrumental errors. These are caused by faulty experimental equipment such as uncalibrated electronic balance and volumetric flask.

1.1.3 Errors of materials. These are resulted from unqualified reagents and distilled water with impurities.

1.1.4 Operative errors. These arise from the operator's personal reason or operation habit. For example, volume readings of burette from some people are always high, while those from others are always low. For the color judgement of the solution at the endpoint in a titration, some people tend to get a light color solution, while some are inclined to obtain a dark color solution.

Systematic errors can be eliminated via correction based on the nature of them.

1.2 Random errors

Random errors cannot be controlled and determined and they are often called indeterminate errors or accidental errors, which occur in any measurement. Sources of accidental errors include changes in instrument performance, fluctuations of environmental temperature, humidity and atmospheric pressure, and small differences in operation for parallel samples. Although both the direction and magnitude of indeterminate errors are random, they conform to statistic rules: in a series of measurements, there should be few large errors and numbers of positive and negative errors are supposed to be equal.

It is almost impossible to eliminate all possible accidental errors, nevertheless, they can be minimized to a tolerable level through multiple measurements. What is noteworthy is that systematic errors and accidental errors always occur together in an experiment.

2. Accuracy and precision

2.1 Accuracy and errors

Accuracy is the degree of agreement between the measured value and the true value. The closer the measured value is to the true value is, the more accurate the result is. Errors can be utilized to measure the degree of accuracy and lower errors indicate more accurate results. Errors can be expressed as absolute error (绝对误差) and relative error (相对误差).

2.1.1 Absolute error (E). It is the difference between the measured value (x) and the true value (μ).

$$E = x - \mu \tag{2-1}$$

Absolute error carries the unit of and its value could be positive or negative. Small absolute error suggests small difference between the measured value and true value, which indicates high accuracy.

2.1.2 Relative error (E_r). It is calculated via dividing the absolute value by the true value and usually is reported as parts per hundred (%). Relative error has no unit and its value could be positive or negative.

$$E_r = \frac{E}{\mu} \times 100\% \tag{2-2}$$

【Example 2-1】

Sample 1 and sample 2 were weighed with an analytical balance. The measured value and the true value of sample 1 are 0.5505g and 0.5504g respectively, and those of sample 2 are 0.1505g and 0.1504g respectively. Absolute errors for both of the two samples are the same, namely 0.0001g, what are the relative errors for them?

Solution:

$$E_r(1) = \frac{+0.0001}{0.5504} \times 100\% = +0.02\%$$

$$E_r(2) = \frac{+0.0001}{0.1504} \times 100\% = +0.07\%$$

The above results indicate that the bigger the absolute value, the smaller the relative error and the higher the accuracy when absolute values are identical. It is more useful to use E_r to evaluate the accuracy of the analysis results.

2.1.3 True value. An absolute true value is seldom known, so the accepted true value is used instead for the definition of accuracy. Some common true values used in analytical chemistry are:

Theoretical true value (理论真值). For example, the theoretical composition of compounds and the angles of a triangle add up to 180 degrees.

Conventional true value (约定真值). It includes international units defined by General Conference of Weights and Measures (CGPM) and China statutory measurement units, such as mole and relative atomic weight.

Relative true value (相对真值). It is the measured value of a known standard sample. After the multiple measured results from different laboratories and operators are handled with mathematical statistics, the relatively accurate contents of each component are obtained and they are called relative true values or standard values.

2.2 Precision and deviation

Precision describes the reproducibility of measurements: in other words, the closeness of results that have been obtained in exactly the same way. The closer the measured values are, the higher the precision of the measurement is. On the contrary, the more scattered the measured values, the lower the precision of the measurement. The precision can be expressed as the standard deviation. The smaller the deviation, the higher the precision. Deviations include:

2.2.1 Deviation (绝对偏差, d). It is the difference between a individual value (x_i) and the mean value ($\bar{x}$) of the measurement. It carries the unit of x_i and its value could be positive or negative.

$$d = x_i - \bar{x} \tag{2-3}$$

2.2.2 Average deviation (平均偏差, $\bar{d}$). It is the average of the absolute value of each deviation. It carries the unit of x_i and its value is positive.

$$\bar{d} = \frac{\sum_{i=1}^{n}|x_i - \bar{x}|}{n} \tag{2-4}$$

where n represents the number replicate measurements.

2.2.3 Relative average deviation (相对平均偏差, $\bar{d}_r$). It is calculated via dividing the average deviation by the average value and usually is reported as parts per hundred (%).

$$\bar{d}_r\% = \frac{\bar{d}}{\bar{x}} \times 100\% \tag{2-5}$$

Though it is simple to use average deviation and relative average deviation to express precision, the average deviation calculated by all measurements will be small because that measurements with large error are always few. Therefore, standard deviation and relative standard deviation are used more often.

2.2.4 Standard deviation (标准偏差, s). When the number of measurements $n \leqslant 20$, the standard deviation can be used to indicate the dispersion degree of the measured values. Standard deviation is given by the following equation and has the same unit as x_i.

$$s = \sqrt{\frac{\sum_{i=1}^{n}(x_i - \bar{x})^2}{n-1}} \tag{2-6}$$

2.2.5 Relative standard deviation (相对标准偏差, RSD). It is the ratio of the standard deviation to the mean value and usually is reported as parts per hundred (%).

$$\text{RSD} = \frac{s}{\bar{x}} \times 100\% \tag{2-7}$$

Among all deviations, the relative standard deviation is used more often because of higher reliability.

【Example 2-2】

Results from five replicate determinations of $CaCO_3$ in tap water are 135.6, 136.0, 136.5, 135.8 and 136.1mg/L. Calculate the average value, average deviation, relative average deviation, standard deviation and relative standard deviation.

Solution:

$$\bar{x} = \frac{135.6 + 136.0 + 136.5 + 135.8 + 136.1}{5} = 136.0 \text{ (mg/L)}$$

$$\bar{d} = \frac{|135.6-136.0| + |136.0-136.0| + |136.5-136.0| + |135.8-136.0| + |136.1-136.0|}{5}$$

$$= 0.24 \text{ (mg/L)}$$

$$\bar{d}_r\% = \frac{\bar{d}}{\bar{x}} \times 100\% = \frac{0.24}{136.0} \times 100\% = 0.18\%$$

$$s = \sqrt{\frac{\sum_{i=1}^{n}(x_i - \bar{x})^2}{n-1}} = \sqrt{\frac{0.4^2+0.5^2+0.2^2+0.1^2}{5-1}} = 0.34 \text{(g/L)}$$

$$\text{RSD} = \frac{s}{\bar{x}} \times 100\% = \frac{0.34}{136.0} \times 100\% = 0.25\%$$

2.3 The relationship between accuracy and precision

Although there is a conceptual distinction between accuracy and precision, they are closely related to each other. For example, four experimenters respectively tested the same sample in sextuplicate, and the true value was 10.00%, as shown in Figure 2-1.

The good precision and poor accuracy of A indicate a small accidental error and a large systematic error. The excellent precision and accuracy of B demonstrate small accidental error and systematic error. The average value of C is close to the true value by accident because the positive errors and negative errors cancel each other out. For the result of D, the low precision and accuracy suggest large accidental error and systematic error.

The systematic error affects the accuracy of the analysis results, while the random error affects the precision. The high precision does not mean high accuracy because there may be systematic error; the low precision indicates unreliable result and then makes no sense to consider the accuracy. Hence,

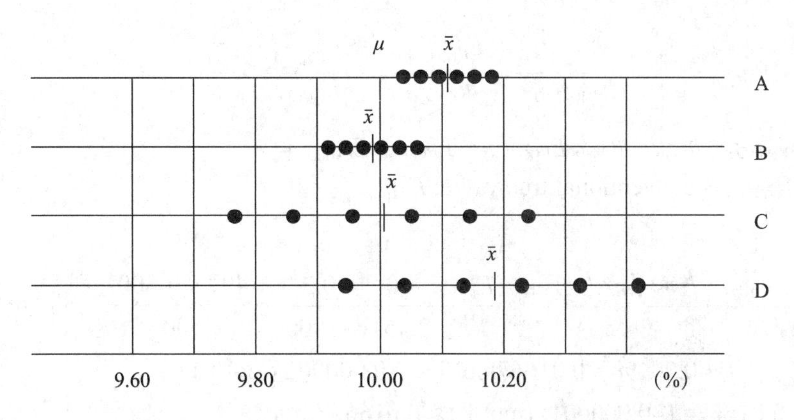

Figure 2-1 Accuracy and precision in quantitative analysis

high precision is the prerequisite to high accuracy. On the premise that systematic errors are eliminated, precision can be used to express the accuracy of the measurement results.

3. Propagation of error

For most experiments, we need to perform arithmetic operations on several numbers, each of which has an error and the errors in the individual numbers will propagate to the result via the calculations. The rules for propagation of systematic error and random error are different and it is summarized in Table 2-1.

Table 2-1 Summary of rules for propagation of errors

Function	Systematic error	Random error
1. $R=x+y-z$	$E_R=E_x+E_y-E_z$	$s_R^2=s_x^2+s_y^2+s_z^2$
2. $R=x\cdot y/z$	$\frac{E_R}{R}=\frac{E_x}{x}+\frac{E_y}{y}-\frac{E_z}{z}$	$\left(\frac{s_R}{R}\right)^2=\left(\frac{s_x}{x}\right)^2+\left(\frac{s_y}{y}\right)^2+\left(\frac{s_z}{z}\right)^2$

3.1 Propagation of systematic errors

The rules for propagation of the systematic error are: for addition and subtraction, the absolute error in the answer is the sum and difference of the absolute errors of the individual terms; for multiplication and division, the relative error in the answer is the sum and difference of the relative errors of the individual terms, as shown in Table 2-1.

【Example 2-3】

You prepared a 0.01667mol/L $K_2Cr_2O_7$ solution by dissolving 2.4516g $K_2Cr_2O_7$ and diluting the above solution up to 500ml. The absolute error of the initial and final readings of analytical balance is respectively −0.2mg and + 0.3mg, and the actual volume of volumetric flask is 499.93ml. Find the relative error, absolute error and actual concentration of the $K_2Cr_2O_7$ solution.

Solution:

Based on

$$c_{K_2Cr_2O_7}=\frac{m}{M_{K_2Cr_2O_7}\times V}\ (\text{mol/L})$$

The relative error of the result is

$$\frac{E_{c_{K_2Cr_2O_7}}}{c_{K_2Cr_2O_7}}=\frac{E_{m_{K_2Cr_2O_7}}}{m_{K_2Cr_2O_7}}-\frac{E_{M_{K_2Cr_2O_7}}}{M_{K_2Cr_2O_7}}-\frac{E_V}{V}$$

Because $m_{K_2Cr_2O_7}=m_{initial}-m_{final}$, $E_{m_{K_2Cr_2O_7}}=E_{m_{initial}}-E_{m_{final}}$

Because $M_{K_2Cr_2O_7}$ is conventional true value, $E_{M_{K_2Cr_2O_7}}=0$

Then

$$\frac{E_{c_{K_2Cr_2O_7}}}{c_{K_2Cr_2O_7}}=\frac{E_{m_{before}}-E_{m_{after}}}{m_{K_2Cr_2O_7}}-\frac{E_V}{V}=\frac{-0.2-0.3}{2.4516\times1000}-\frac{499.93-500}{500}\approx-0.006\%$$

$$E_{c_{K_2Cr_2O_7}}=-0.006\%\times0.01667\text{mol/L}=-0.000001\ (\text{mol/L})$$

$$c=0.01667-(-0.000001)\ (\text{mol/L})\approx0.01667\ (\text{mol/L})$$

3.2 Propagation of random errors

The most likely uncertainty in the result is not simply the sum of the individual errors, because some of them are likely to be positive and some negative. The rules for propagation of random error are: for addition and subtraction, the absolute variance of the answer is the sum of the individual variances; for multiplication or division, the relative variance of the answer is the sum of the individual relative variances, as shown in Table 2-1.

【Example 2-4】

The mass weighed by an analytical balance is the difference between final and initial readings. If the standard deviation s in each reading of analytical balance is 0.10mg, what is the standard deviation s_m in the mass weighed?

Solution:

Suppose that the initial reading is m_1 and the final reading is m_2, the mass weighed m is the difference:

$$m=m_1-m_2$$

The standard deviation s_m in the mass weighed is:

$$s_m=\sqrt{s_1^2+s_2^2}=\sqrt{2s^2}=0.14(\text{mg})$$

PPT

Section 2 Significant Figures

Significant Figures are the digits in a measured quantity, including all digits known exactly and one digit (the last) whose quantity is uncertain. Recording a measurement provides information about both its magnitude and uncertainty. For example, if we weigh a sample on a balance and record its mass as 1.2637g, we assume that all digits, except the last, are known exactly. We assume that the last digit has an uncertainty of at least ±1, giving an absolute uncertainty (绝对误差) of at least ±0.0001g, or a relative uncertainty (相对误差) of at least:

$$\frac{\pm0.0001\text{g}}{1.2637\text{g}}\times100\%=\pm0.0079\%$$

1. Counting rules

A significant figure is a digit which denotes the amount of the quantity in the place in which it stands.

The digit zero is a significant figure except when it is the first figure in a number. Thus, in the quantities 1.2680g and 1.0062g the zero is significant, but in the quantity 0.0025kg the zeros are not significant figures; they serve only to locate the decimal point and can be omitted by proper choice of units, i.e. 2.5g. The first two numbers contain five significant figures, but 0.0025 contains only two significant figures.

The number 6.302×10^{-6} has four significant figures, because all four digits are necessary. You could write the same number as 0.000006302, which also has just four significant figures. The zeros to the left of the 6 are merely holding decimal places. The number 92 500 is ambiguous. It could mean any of the following:

9.25×10^{4} 3 significant figures

9.250×10^{4} 4 significant figures

9.2500×10^{4} 5 significant figures

You should write one of the three numbers above, instead of 92 500, to indicate how many figures are actually known.

For measurements using logarithms, such as pH, the number of significant figures is equal to the number of digits to the right of the decimal, including all zeros. Digits to the left of the decimal are not included as significant figures since they only indicate the power of 10. A pH of 2.45, therefore, contains two significant figures.

Exact numbers, such as the stoichiometric coefficients (化学计量系数) in a chemical formula or reaction, and unit conversion factors, have an infinite number of significant figures. A mole of $CaCl_2$, for example, contains exactly two moles of chloride and one mole of calcium. In the equality 1000ml = 1L, both numbers have an infinite number of significant figures.

2. Significant figures in arithmetic

We now consider how many digits to retain in the answer after you have performed arithmetic operations with your data. Rounding should only be done on the final answer (not intermediate results), to avoid accumulating round-off errors.

2.1 Rounding off

There are several different rules commonly used for discarding unwanted digits in a number. The following rule, which is used in this book, is the simplest and most common.

(1) If the digit to be discarded is more than 5, increase the retained preceding digit by 1. Thus, rounding 12.486 to three digits gives 12.5.

(2) If the digit to be discarded is less than 5, do not change the preceeding digit. Thus, rounding 12.442 to three digits gives 12.4.

(3) If the digit to be discarded is exactly 5, then round the least significant figure to the nearest even number; thus, rounding 12.450 to the nearest tenth gives 12.4, but rounding 12.550 to the nearest tenth gives 12.6. Rounding in this manner prevents us from introducing a bias by always rounding up or down.

Finally, to avoid "round-off" errors in calculations, it is a good idea to retain at least one extra significant figure throughout the calculation. This is the practice adopted in this textbook. Better yet, invest in a good scientific calculator that allows you to perform lengthy calculations without recording intermediate values. When the calculation is complete, the final answer can be rounded to the correct number of significant figures.

2.2 Addition and subtraction

If the numbers to be added or subtracted have equal numbers of digits, the answer goes to the same decimal place as in any of the individual numbers:

$$\begin{array}{r} 1.362\times10^{-4} \\ +\ \ 3.111\times10^{-4} \\ \hline 4.473\times10^{-4} \end{array}$$

If the numbers being added do not have the same number of significant figures, we are limited by the least certain one. For example, the molecular mass of KrF_2 is known only to the third decimal place, because we know the atomic mass of Kr to only three decimal places:

$$\begin{array}{rl} 18.998\,4032 & \text{(F)} \\ +\ \ 18.998\,4032 & \text{(F)} \\ +\ \ 83.798 & \text{(Kr)} \\ \hline 121.794\,8064 & \end{array}$$

The number 121.794 806 4 should be rounded to 121.795 as the final answer.

2.3 Multiplication and division

In multiplication and division, we are normally limited to the number of digits contained in the number with the fewest significant figures.

$$\frac{22.91\times0.152}{16.302}=0.21361=0.214$$

It is important to remember, however, that these rules are generalizations. What is conserved is not the number of significant figures, but absolute uncertainty when adding or subtracting, and relative uncertainty when multiplying or dividing. For example, the following calculation reports the answer to the correct number of significant figures, even though it violates the general rules outlined earlier.

$$\frac{101}{99}=1.02$$

Since the relative uncertainty in both measurements is roughly 1% (101±1, 99±1), the relative uncertainty in the final answer also must be roughly 1%. Reporting the answer to only two significant figures (1.0), as required by the general rules, implies a relative uncertainty of 10%. The correct answer, with three significant figures, yields the expected relative uncertainty.

PPT

Section 3 Statistical Treatment of Analytical Data

Experimental measurements always contain some variability, so no conclusion can be drawn with certainty. Statistics gives us tools to accept conclusions that have a high probability of being correct and to reject conclusions that do not.

1. The normal distribution of random errors

We now know that random variations will be present whenever we make a measurement. If we were

to make the same measurement many times and plot the number of times we obtain any particular value, we would get a result similar to the plot in Figure 2-2. In this graph, the x-axis gives the range of values we have measured for our sample, and the y-axis shows the number of times each of these values was obtained. If we make enough measurements, the plot we get will have a "bell shape" with the center occurring at the average of our data set. This is known as a normal distribution (正态分布) or a Gaussian distribution (高斯分布).

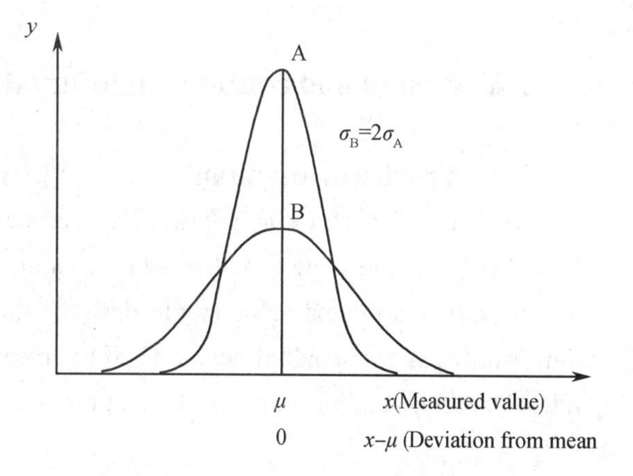

Figure 2-2 The normal distribution curve

The shape of the normal distribution curve (正态分布曲线) can be described by the following equation:

$$y = f(x) = \frac{1}{\sigma\sqrt{2\pi}} e^{-(x-\mu)^2/2\sigma^2} \tag{2-8}$$

Along with x and y, two other factors that appear in this equation are (1) the population mean μ (总体平均值), which gives the central point for the distribution, and (2) the population standard deviation σ (总体标准差), which describes the width of this curve. As we increase or decrease μ, the entire curve shifts to higher or lower x values. As we increase or decrease σ, the curve becomes broader or narrower, respectively. The two curves in Figure 2-2 are for two populations of data that differ only in their σ. The σ for the data set yielding the broader but lower curve B is twice that for the measurements yielding curve A. The breadth of these curves is a measure of the precision of the two sets of data. Thus, the precision of the data set leading to curve A is twice as good as that of the data set represented by curve B.

A normal distribution curve has several general properties: (a) The mean occurs at the central point of maximum frequency, (b) there is a symmetrical distribution of positive and negative deviations about the maximum, and (c) there is an exponential decrease in frequency as the magnitude of the deviations increases. Thus, small uncertainties are observed much more often than very large ones.

In order to be convenient, a new variable u is introduced, defined as:

$$u = \frac{x - \mu}{\sigma} \tag{2-9}$$

Since u is the deviation from the mean relative to the standard deviation, a plot of relative frequency versus u yields a single normal distribution curve, called a standard normal distribution curve (标准正态分布曲线), which describes all populations of data regardless of standard deviation. Thus, Figure 2-3 is the normal error curve for both sets of data used to plot curves A and B in Figure 2-2.

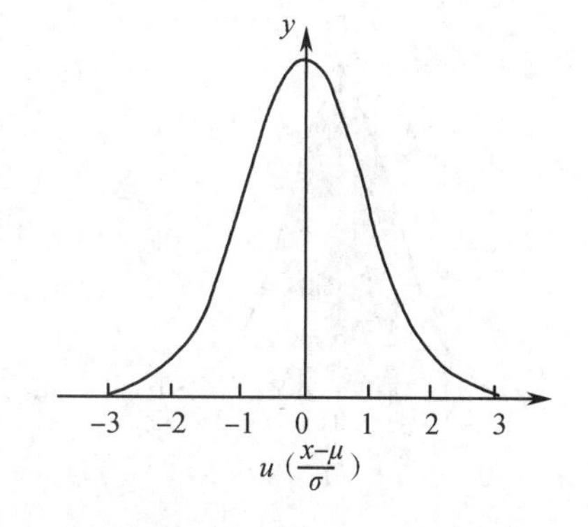

Figure 2-3 The standard normal distribution curve

2. Precision and confidence interval of mean

2.1 Precision of mean

Precision of mean (平均值的精密度) can be expressed by standard deviation of the mean (平均值的标准偏差). In the same way that we use standard deviation s to describe the variation within a data set, we can employ a related value ($s_{\bar{x}}$) to describe the precision of our experimental average ($\bar{x}$). This new value, known as the standard deviation of the mean, is determined by using the standard deviation of the entire data set (s) and the number of data points in this set (n).

$$s_{\bar{x}} = \frac{s}{\sqrt{n}} \tag{2-10}$$

Equation 2-10 tells us that the mean of four measurements is more precise by $\sqrt{4} = 2$ than individual measurements in the data set. For this reason, averaging results is often used to improve precision. This effect is illustrated in Figure 2-4 and occurs because the precision of the experimental average becomes better as we acquire more data, making $\bar{x}$ a more reliable estimate of the true average. Although using more measurements will provide a better estimate of the mean result for a sample by decreasing $s_{\bar{x}}$, acquiring more data will also increase the time, effort, and sample required to make the measurements. As a compromise between effort and reproducibility, three to five measurements are recommended for most analyses.

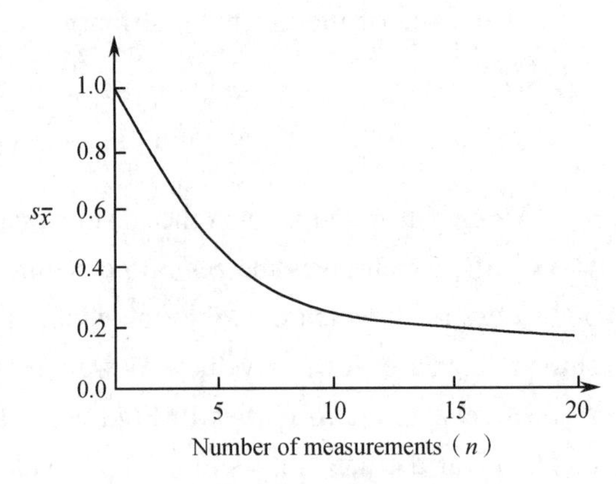

Figure 2-4 The relationship between $s_{\bar{x}}$ and number of measurements (n)

2.2 Confidence intervals

Figure 2-5 shows a series of three standard normal distribution curves. In each, the relative frequency is plotted as a function of the quantity u. The numbers within the shaded areas are the percentage of the total area under the curve. For example, as shown in curve (a), 68.3% of the area under any normal distribution curve is located between -1σ and $+1\sigma$. Proceeding to curves (b) and (c), we see that 90% of the total area lies between -1.64σ and $+1.64\sigma$ and 95% between -1.96σ and $+1.96\sigma$. Relationships such as these allow us to define a range of values around a measurement result within which the true mean is likely to lie with a certain probability provided we have a reasonable estimate of σ. For example, if we

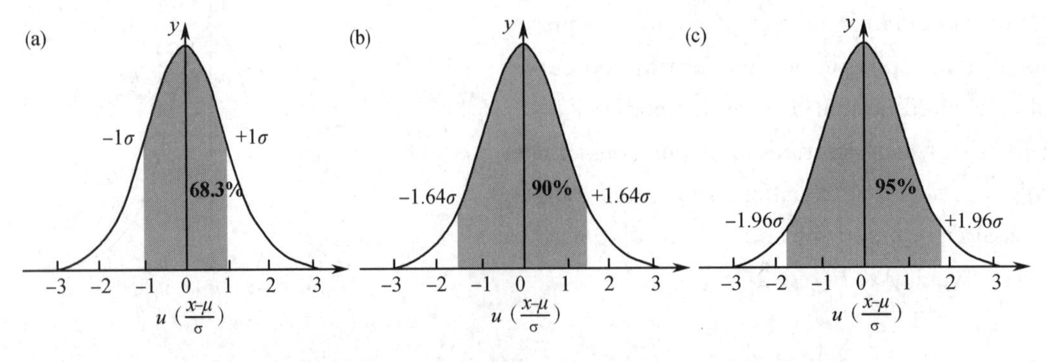

Figure 2-5 Areas under a normal distribution curve for various values of $\pm u$.

have a result x from a data set with a standard deviation of σ, we may assume that 90 times out of 100 the true mean μ will fall in the interval $x\pm1.64\sigma$ (see Figure 2-5b). The probability is called the confidence level (置信度, P). In the example of Figure 2-5b, the confidence level is 90% and the confidence interval (置信区间) is from -1.64σ to $+1.64\sigma$. The probability that a result is outside the confidence interval is often called the significance level (显著性水平, α), where $\alpha=1-P$.

If we make a single measurement x from a distribution of known σ, we can say that the true mean should lie in the interval $x \pm u\sigma$ with a probability dependent on u. Thus,

$$\mu = x \pm u\sigma \tag{2-11}$$

However, we rarely estimate the true mean from a single measurement. Instead, we use the experimental mean $\overline{x}$ of n measurements as a better estimate of μ, that is,

$$\mu = \overline{x} \pm \frac{u\sigma}{\sqrt{n}} \tag{2-12}$$

From a limited number of measurements, we cannot find the true standard deviation, σ. What we determine is S, the sample standard deviation. As indicated earlier, s calculated from a small set of data may be quite uncertain. To account for the variability of s, we use the important statistical parameter t, which is defined in exactly the same way as u (Equation 2-9) except that s is substituted for σ. For a single measurement with result x, we can define t as:

$$t = \frac{x-\mu}{s} \tag{2-13}$$

For the mean of n measurements,

$$t = \frac{x-\mu}{\frac{s}{\sqrt{n}}} \tag{2-14}$$

The confidence interval for the mean of n replicate measurements can be calculated from t by Equation 2-15, which is similar to Equation 2-15 using u:

$$\mu = \overline{x} \pm \frac{ts}{\sqrt{n}} \tag{2-15}$$

Table 2-2 gives the values for t. The t value depends on the number of degrees of freedom (自由度 , f), where $f = n-1$. As you add more data points, the value of t decreases and approaches a constant. This reflects the fact that the experimental mean and standard deviations are becoming better estimates of their true values. Another factor that will determine the selected value of t is the degree of certainty you would like to have that the true answer falls within your calculated confidence interval. This degree of certainty is known as the confidence level. As you go to higher confidence levels, the size of t increases to provide a greater chance for the true value falling within the stated range.

Table 2-2 Values of *t* at various degrees of freedom

Degrees of freedom (f)	Confidence level (P)		
	90%	95%	99%
1	6.31	12.71	63.66
2	2.92	4.30	9.92
3	2.35	3.18	5.84

continued

Degrees of freedom (f)	Confidence level (P)		
	90%	95%	99%
4	2.13	2.78	4.60
5	2.02	2.57	4.03
6	1.94	2.45	3.71
7	1.90	2.36	3.50
8	1.86	2.31	3.36
9	1.83	2.26	3.25
10	1.81	2.23	3.17
20	1.72	2.09	2.84
∞	1.64	1.96	2.58

【Example 2-5】

The mean of four determinations of the copper content of a sample of an alloy was 8.27 wt% with a standard deviation $s = 0.17$. Calculate the confidence interval at the 95% confidence level.

Solution: For the 94% confidence interval, look up t in Table 2-2 under 95 and across from three degrees of freedom. The value of t is 3.18, so the 95% confidence interval is:

$$95\% \text{ confidence interval} = \bar{x} \pm \frac{ts}{\sqrt{n}} = 8.27 \pm \frac{3.18 \times 0.17}{\sqrt{4}} = (8.27 \pm 0.27) \text{ wt\%}$$

Thus, there is 95% confidence that the true value of the copper content of the alloy lies in the range 8.00% to 8.54%.

3. Tests of significance

In developing a new analytical method, it is often desirable to compare the results of that method with true value or those of an accepted (perhaps standard) method. These comparisons make it possible to determine whether the analytical procedure has been accurate and/or precise, or if it is superior to another method.

Deciding whether one set of results is significantly different (显著性差异) from another depends not only on the difference in the means but also on the amount of data available and the spread. There are two common methods for comparing results: the F-test (F 检验) evaluates differences between the spread of results, while the t-test (t 检验) looks at differences between means.

3.1 The F test

This is a test designed to indicate whether there is a significant difference on precision between two methods based on their standard deviations. F is defined in terms of the variances of the two methods, where the variance is the square of the standard deviation (方差 , s^2):

$$F = \frac{s_1^2}{s_2^2} \quad (\text{where } s_1 \geqslant s_2) \tag{2-16}$$

Because we have selected s_2 to be the smaller of the two standard deviations, the value we calculate for F should always be greater than or equal to one. There are two different degrees of freedom, f_1 and f_2, where degrees of freedom (f) is defined as $n-1$ for each case.

If the calculated F value from Equation 2-16 exceeds a tabulated F value at the selected confidence level, then there is a significant difference between the variances of the two methods. On the contrary, if we find that $F_{calculated} \leqslant F_{table}$, the precision of these methods is considered to be the same at the selected confidence level. A list of F values at the 95% confidence level is given in Table 2-3.

Table 2-3 Values of F at the 95% confidence level

Degrees of freedom (f_2)	Degrees of freedom (f_1)									
	2	3	4	5	6	7	8	9	10	∞
2	19.00	19.16	19.25	19.30	19.33	19.36	19.37	19.38	19.39	19.50
3	9.55	9.28	9.12	9.01	8.94	8.88	8.84	8.81	8.78	8.53
4	6.94	6.59	6.39	6.26	6.16	6.09	6.04	6.00	5.96	5.63
5	5.79	5.41	5.19	5.05	4.95	4.88	4.82	4.78	4.74	4.36
6	5.14	4.76	4.53	4.39	4.28	4.21	4.15	4.10	4.06	3.67
7	4.74	4.35	4.12	3.97	3.87	3.79	3.73	3.68	3.63	3.23
8	4.46	4.07	3.84	3.69	3.58	3.50	3.44	3.39	3.34	2.93
9	4.26	3.86	3.63	3.48	3.37	3.29	3.23	3.18	3.13	2.71
10	4.10	3.71	3.48	3.33	3.22	3.14	3.07	3.02	2.97	2.54
∞	3.00	2.60	2.37	2.21	2.10	2.01	1.94	1.88	1.83	1.00

Note: f_1 is degrees of freedom in denominator (s_1); f_2 is degrees of freedom in numerator (s_2).

【Example 2-6】

You are developing a new colorimetric procedure for determining the glucose content of blood serum. You have chosen the standard Folin-Wu procedure with which to compare your results. The standard deviation from the new method is s_1=0.64 (n_A=8 measurements) and the standard deviation from the standard method is s_2=0.41 (n_B=7 measurements). Is there any significant difference between the precision of the two methods?

Solution: To answer the question, find F with Equation 2-16:

$$F = \frac{s_1^2}{s_2^2} = \frac{0.64^2}{0.41^2} = 2.44$$

In looking at Table 2-3, we see that the critical value at the 95% confidence level and for our particular degrees of freedom, where f_1= (n_1-1) =7 and f_2= (n_2-1) = 6, is 4.21. Since the calculated value of 2.44 is less than this, we conclude that there is no significant difference in the precision of the two methods, that is, the standard deviations are from random error alone and do not depend on the sample. In short, statistically, your method does as well as the established procedure.

3.2 The t test

3.2.1 Comparing a mean with a true value

When you work in an analytical laboratory you will often have to compare an experimental result with a known reference value. For instance, this type of comparison might be needed when you are determining the accuracy of a new method. If the reference value is known exactly (or at least has a much better precision than the experimental result), we can use this value to represent the true mean (μ) for the

sample. To compare this value to the average result measured for the sample ($\overline{x}$), we can use the t-value as the test statistic. The result is a statistical method known as the t-test.

The value of t is obtained from the equation:

$$t = \frac{|\overline{x} - \mu|}{s}\sqrt{n} \tag{2-17}$$

Once we have calculated t for our data, we need to compare this result to a critical t value, which is determined by the confidence level (P) and the degrees of freedom (f) for the sample. as obtained from Table 2-2. If we find that $t_{calculated} \leqslant t_{table}$, we can say that $\overline{x}$ and μ are not significantly different at the given confidence level.

【Example 2-7】

Before determining the amount of Na_2CO_3 in an unknown sample, a student decides to check her procedure by analyzing a standard sample known to contain 98.76% *w/w* Na_2CO_3. Five replicate determinations of the standard sample were made with the following results: 98.71%, 98.59%, 98.62%, 98.44%, 98.58%. Is there a significant difference between the mean of the results and the certified or "true" value?

Solution: The mean and standard deviation for the five trials are:

$$\overline{x} = 98.59 \quad s = 0.0973$$

Then we can place these values into Equation 2-17 to calculate a t value.

$$t = \frac{|\overline{x} - \mu|}{s}\sqrt{n} = \frac{|98.59 - 98.76|}{0.0973} \times \sqrt{5} = 3.91$$

The critical t value, as found in Table 2-2, is 2.78 at the 95% confidence level and for $f = (5-1) = 4$ degrees of freedom. Since the experimental t value (3.91) is greater than t_{table}, the conclusion is that a significant difference does exist.

3.2.2 Comparison of Two Experimental Means

Another situation often encountered in chemical analysis is when we need to compare two experimental results. To illustrate this, suppose we have mean results for two samples ($\overline{x}_1$ and $\overline{x}_2$) that have been measured by the same method or by two methods with similar precision. The test statistic we use for this situation is again the t value, but we now need to modify this approach to allow for the fact that both our experimental result and model have some uncertainty in their values. Thus, instead of using the standard deviations for either of these means, we use a pooled standard deviation (合并标准偏差 , s_R) that reflects the variation in both results:

$$s_R = \sqrt{\frac{\sum_{i=1}^{n_1}(x_{1i}-\overline{x}_1)^2 + \sum_{i=1}^{n_2}(x_{2i}-\overline{x}_2)^2}{(n_1-1)+(n_2-1)}} = \sqrt{\frac{s_1^2(n_1-1) + s_2^2(n_2-1)}{(n_1-1)+(n_2-1)}} \tag{2-18}$$

In this equation, s_1 and s_2 are the estimated standard deviations for the two data sets, and n_1 and n_2 are the number of points in each of these sets. You can think of s_R as a weighted average of the individual standard deviations for the two groups of results.

The test statistic t is now found from:

$$t = \frac{|\overline{x}_1 - \overline{x}_2|}{s_R}\sqrt{\frac{n_1 n_2}{n_1+n_2}} \tag{2-19}$$

Once we have calculated this t value, we must compare it to a critical value from Table 2-2, as obtained for our selected confidence level and using $f = (n_1+n_2-2)$ as the degrees of freedom. If the calculated t is less than or equal to t_{table}, we can say that x_1 and x_2 represent the same value at our selected confidence level.

【Example 2-8】

The content of ginsenosides in a Ginseng sample was determined by two different methods. For the first method, $\bar{x}_1$=0.22% with a standard deviation s_1=0.032% (n_1=4 measurements). For the second method, $\bar{x}_2$=0.26% with a standard deviation s_2=0.025% (n_2=3 measurements). Determine whether the difference in their mean values is significant at the 95% confidence level.

Solution: 1. The F test:

$$F=\frac{s_1^2}{s_2^2}=\frac{0.032^2}{0.025^2}=1.64$$

This is less than the tabulated value (9.55), so the two methods have comparable standard deviations and the t test can be applied.

2. The t test:

With Equation 2-18, the pooled standard deviation was calculated, s_R=0.0281

Then calculate t value with Equation 2-19:

$$t=\frac{|\bar{x}_1-\bar{x}_2|}{s_R}\sqrt{\frac{n_1n_2}{n_1+n_2}}=\frac{|0.22-0.26|}{0.0281}\sqrt{\frac{4\times3}{4+3}}=1.87$$

The degrees of freedom in this case is f= (4+3–2) = 5. The tabulated t for five degrees of freedom at the 95% confidence level is 2.57, so there is no statistical difference in the results by the two methods.

4. Rejecting outliers

When a series of replicate analyses is performed, it is not uncommon that one of the results will appear to differ markedly from the others. It is generally considered inappropriate and in some cases unethical to discard data without a reason. However, the questionable result, called an outlier (可疑值), could be the result of an undetected gross error (过失误差). Hence, it is important to develop a criterion to decide whether to retain or reject the outlying data point.

A wide variety of statistical tests have been suggested and used to determine whether an observation should be rejected. Here we introduce the Q test and Grubbs test for discordant values.

4.1 The Q test

The Q test is applied as follows:

(1) Rank the results in the data set in order from the lowest to highest value so that the suspected outlier is either the first or the last data point.

(2) Calculate the difference between the maximum and minimum values: $x_{max}-x_{min}$

(3) Calculate the absolute value of the difference between the questionable result (x_q) and its nearest neighbor (x_n): $|x_q-x_n|$

(4) Calculate the test statistic Q, defined as:

$$Q=\frac{|x_q-x_n|}{x_{max}-x_{min}} \tag{2-20}$$

(5) Compare the calculated Q value to a critical test value, as given in Table 2-4. If $Q_{calculated}\leqslant Q_{table}$, the suspected outlier must be retained. If $Q_{calculated}>Q_{table}$, the outlier may be rejected.

Table 2-4 Critical values of Q

Number of observations (n)	Confidence level (P)		
	90%	95%	99%
3	0.94	0.97	0.99
4	0.76	0.84	0.93
5	0.64	0.73	0.82
6	0.56	0.64	0.74
7	0.51	0.59	0.68
8	0.47	0.54	0.63
9	0.44	0.51	0.60
10	0.41	0.49	0.57

4.2 The G (Grubbs) test

The G test is applied as follows:

(1) Calculate the average ($\bar{x}$) and the standard deviation s of the complete data set.

(2) Calculate the absolute value of the difference between the questionable result (x_q) and the mean value: $|x_q - \bar{x}|$

(3) Calculate the Grubbs statistic G, defined as:

$$G = \frac{|x_q - \bar{x}|}{s} \tag{2-21}$$

(4) Compare the calculated G value to a critical test value, as given in Table 2-5. If $G_{calculated} \leqslant G_{table}$, the suspected outlier must be retained. If $G_{calculated} > G_{table}$, the outlier may be rejected.

In Grubbs test, not only set up certain confidence level, but also introduce the mean and standard deviation, so its accuracy is better than that of Q test.

Table 2-5 Critical values of G

Number of observations (n)	Confidence level (P)		
	95%	97.5%	99%
3	1.15	1.15	1.15
4	1.46	1.48	1.49
5	1.67	1.71	1.75
6	1.82	1.89	1.94
7	1.94	2.02	2.10
8	2.03	2.13	2.22
9	2.11	2.21	2.32
10	2.18	2.29	2.41
11	2.23	2.36	2.48

continued

Number of observations (n)	Confidence level (P)		
	95%	97.5%	99%
12	2.29	2.41	2.55
13	2.33	2.46	2.61
14	2.37	2.51	2.63
15	2.41	2.55	2.71
20	2.56	2.71	2.88

【Example 2-9】

The analysis of a city drinking water for arsenic yielded values of 5.60, 5.64, 5.69, 5.70 and 5.81 mg/L. Use both Q test and G test to determine whether the last value appears anomalous; should it be rejected at the 95% confidence level?

Solution: 1. The Q test:

$$Q=\frac{|x_q-x_n|}{x_{\max}-x_{\min}}=\frac{|5.81-5.70|}{5.81-5.60}=0.52$$

For five measurements, Q_c at the 95% confidence level is 0.73. Because 0.52 < 0.71, we must retain the outlier at the 95% confidence level.

2. The G test:

The mean and standard deviation for the five trials are:

$$\bar{x}=5.69 \qquad s=0.0792$$

$$G=\frac{|x_q-\bar{x}|}{s}=\frac{|5.81-5.69|}{0.0792}=1.51$$

For five measurements, G_c at the 95% confidence level is 1.67. Because 1.51 < 1.67, we must retain the outlier at the 95% confidence level.

5. Correlation and regression

For most chemical analyses, the response of the procedure must be evaluated for known quantities of analyte (called standards) so that the response to an unknown quantity can be interpreted. For this purpose, we commonly prepare a calibration curve (校正曲线), which is constructed by measuring the analytical signal for each standard and plotting this response against concentration. Provided the same experimental conditions are used for the measurement of the standards and for the test (unknown) sample, the concentration of the latter may be determined from the calibration curve by graphical interpolation.

There are two statistical tests that should be applied to a calibration curve:

(a) to ascertain if the graph is linear, or in the form of a curve;

(b) to evaluate the best straight line (or curve) throughout the data points.

5.1 Correlation coefficient

In order to establish whether there is a linear relationship between two variables x_1 and y_1, the correlation coefficient (相关系数, r) is used:

$$r = \frac{\sum_{i=1}^{n}(x_i-\bar{x})(y_i-\bar{y})}{\sqrt{\sum_{i=1}^{n}(x_i-\bar{x})^2\sum_{i=1}^{n}(y_i-\bar{y})^2}} \tag{2-22}$$

where n is the number of data points, x_1 and y_1 are the individual values of the variables x and y, respectively, $\bar{x}$ and $\bar{y}$ and are their means.

The value of r must lie between +1 and −1: the nearer it is to +1, or in the case of negative correlation to −1, then the greater the probability that a definite linear relationship exists between the variables x and y. Values of r that tend towards zero indicate that x and y are not linearly related (they may be related in a non-linear fashion). A positive value for r means y and x are changing in the same direction (for example, the value of y increases as x increases), while a negative value for r indicates y and x are changing in opposite direction.

5.2 Linear regression

Once a linear relationship has been shown to have a high probability by the value of the correlation coefficient (r), then the best straight line through the data points has to be estimated. This can often be done by visual inspection of the calibration graph but in many cases it is far better practice to evaluate the best straight line by linear regression (线性回归), with the method of least squares (最小二乘法).

Linear regression involves taking a set of (x, y) values and fitting these to an equation with the following form:

$$y = ax + b \tag{2-23}$$

where y is the dependent variable, x is the independent variable, y is usually the measured variable, plotted as a function of changing x. For example, in a spectrophotometric calibration curve, y would represent the measured absorbances and x would be the concentrations of the standards.

The slope of the curve (a) and the intercept on the y axis (b) are given by the following equations:

$$a = \frac{\sum_{i=1}^{n}(x_i-\bar{x})(y_i-\bar{y})}{\sum_{i=1}^{n}(x_i-\bar{x})^2} \tag{2-24}$$

$$b = \bar{y} - a\bar{x} \tag{2-25}$$

where $\bar{x}$ is the mean of all values of x_i, and $\bar{y}$ is the mean of all values of y_i.

PPT

Section 4 Methods to Improve the Accuracy of Analysis Results

Measures could be taken to eliminate systematic errors and reduce accidental errors so as to improve the accuracy of analysis results. Some measures include:

1. Selection of appropriate analysis method

Demands to the accuracy of the analysis results depend on the content of the components. For components with high content, the relative error is generally required to be less than a few thousandths. As for the micro and trace components, the relative error is generally within 1%~5%. Appropriate

analysis methods should be chosen based on specific requirements. Although classical chemical analysis has low sensitivity, it can get accurate results for components with high content, and of which relative error is less than a few thousandths. Instrumental analysis has the advantages of high sensitivity and low absolute error; it can be used for the detection of micro or trace components in spite of large relative error. Therefore, chemical analysis is mainly used for the analysis of major constituents and instrumental analysis is suitable for micro and trace components.

Besides the content of the components, other interfering substances should also be considered. To sum up, when choosing analytical method, we need to take various factors into consideration including sample composition, the content of components to be measured and the requirements on the analysis result.

2. Reduction of measurement errors

Errors exist in all measurements. In order to ensure the accuracy of the analysis results, it is necessary to minimize the error of each analysis step. For example, in the weighing, the absolute error of analytical balance is ±0.0001g; twice balances are needed in one weighing and the maximum error introduced is ±0.0002g. In order to make the relative error of weighing≤0.1%, the sample weight should≥0.2g.

It should be pointed out that requirements for the accuracy of result should be compatible with the accuracy of analytical methods. For instance, when the content of micro components is determined by the colorimetric method with a relative error of 2%, the absolute error of weighing for 0.5g sample is ± 0.5×2% = ± 0.01g, which means that it only needs balance with absolute error of ± 0.01g rather than the balance with absolute error of ±0.0001g. Nevertheless, it is better to use balance with absolute error of ±0.001g in order to make the weighing error negligible.

3. Reduction of random errors

On the premise that systematic errors are eliminated, the greater the number of parallel measurements, the closer the average value is to the true value according to the distribution of random errors. Therefore, random errors can be reduced through multiple parallel measurements and three to four times are enough for a sample in general.

4. Detection and elimination of systematic errors in measurement

4.1 Control experiment

It is the most common and effective method to check the systematic errors. Control experiments (对照实验) can be classified into two categories. One is to analyze the standard sample with known content (standard value) using the selected analysis method. After significance test is conducted on the standard value and the measured result, systematic errors of the method are estimated. The other is to determine the sample by the selected method and the recognized classical method at the same time, and to test the significance of the results to decide the systematic errors of the selected method. The correction value can be calculated based on the results of the control test and it can be used to correct the systematic errors.

4.2 Recovery test

It is used when the composition of the sample is unclear or there is no standard sample for the control experiment. After the analyte in the sample is determined, the sample spiked with known amount of the pure substance or standard of the analyte (sample + spike) is detected by the same procedure. The recovery rate (回收率) is calculated by the following formula:

$$\text{Recovery (\%)} = \frac{\text{measured value of "sample+spike"} - \text{measured value of sample}}{\text{amount spiked into the sample}}$$

The recovery rate has a range of 95% to 105%. The closer the recovery rate is to 100%, the smaller the system error is and the higher the accuracy is.

4.3 Blank determination

It is an analysis on the added reagents only without sample, and the result is expressed as blank value and is applied as a correction to the sample measurement. Blank determination (空白实验) can reduce the systematic error caused by interfering contaminants from the reagents and vessels employed in the analysis. If the blank value is large, the blank value should be reduced by purifying reagents, using qualified solvents or replacing test vessels.

4.4 Calibration of equipment

Errors caused by faulty instruments can be eliminated via the calibration. The performance of equipment may vary with the environment, time and other conditions, in consequence, equipment is supposed to be calibrated termly.

4.5 Checklist of operations

Operative errors can be minimized by having a checklist of operations.

重点小结

分析化学定量分析时，不仅要测得被测组分的含量，还要了解分析过程中误差产生的原因及其规律，采用相应的措施，减小误差，使分析结果尽可能接近真值。分析误差可分为系统误差和偶然误差。其中系统误差是由某种确定的原因造成的误差，会在重复测定中重复出现、方向与大小固定；偶然误差是由某些难以控制的偶然因素引起的误差，方向与大小不定、符合统计规律。系统误差和偶然误差的大小会分别影响分析结果的准确度和精密度。准确度是指测量值与真实值接近的程度，其高低用误差衡量；精密度是指在相同条件下，平行测量的各测量值之间相互接近的程度，其高低用偏差衡量。准确度高要求精密度一定要高，精密度是保证准确度的先决条件；但精密度高，准确度不一定高，因为可能存在系统误差。每个测量值的误差，可通过计算传递到分析结果中。系统误差和偶然误差的传递规律不同，系统误差传递绝对误差，偶然误差传递相对误差。

有效数字是指在分析工作中实际上能测量到的、有实际意义的数字。在记录有效数字时，只允许保留最后一位不准确的数字（有 ±1 个单位误差）和所有准确的数字。采用“四舍六入五留双”的规则对有效数字进行修约。有效数字的运算规则分为加减法和乘除法规则。

在舍弃离群值和消除系统误差后，进行无限次测量，测量数据的偶然误差符合正态分布。进行有限次测量时，偶然误差符合 t 分布，由 t 分布可以求出平均值的置信区间。在分析工作中，首先用 Q 检验法或 G 检验法对离群值进行判断，确定其取舍后进行差别检验，先作 F 检验，确认两组数据的精密度无显著性差异后，进行 t 检验，即准确度的检验。研究变量之间关系可以用

到相关分析与回归分析。相关分析用来评价两变量间相关的程度，回归分析用来建立变量间的数学模型。

目标检测

1. State whether the errors in parts (a) – (e) are random or systematic:

(a) The inconformity between the titration endpoint and stoichiometric point.

(b) Weighing with corroded weights.

(c) Slight solubility of a precipitate in a gravimetric analysis.

(d) Temperature fluctuation during weighing

(e) A buret reading of 23.65ml, is written down as 26.35ml.

2. Explain the difference and relation between systematic and random errors.

3. How to check the systematic error in the analysis result?

4. Explain the relation between accuracy and precision.

5. To deliver a certain volume from a buret requires two readings: initial and final. If each reading is made with a standard deviation of 0.02ml, what is the standard deviation in the volume delivered?

6. How to improve the accuracy of results?

7. Find the average, average deviation, relative average deviation, standard deviation and relative standard deviation.

(1) 33.45, 33.49, 33.40, 33.46

(2) 0.1046, 0.1043, 0.1039, 0.1044

8. How many significant figures are there in the following numbers?

(a) 5.0330 (b) 0.05600 (c) 1.60×10^4

9. Round each number as indicated:

(a) 1.2367 to 4 significant figures

(b) 1.2384 to 4 significant figures

(c) 0.1352 to 3 significant figures

(d) 2.051 to 2 significant figures

(e) 2.0050 to 3 significant figures

10. Write each answer with the correct number of significant figures.

(a) $34.021+2.88 =$

(b) $12.133-8.63=$

(c) $4.342\times9.21 =$

(d) $0.0702\div(5.783\times10^3) =$

(e) $\lg(3.098\times10^5) =$

11. What are the main factors that determine the shape of a normal distribution? How does the shape of this distribution change as each of these factors is altered?

12. What is the meaning of a confidence interval? Discuss how the size of the confidence interval for the mean is influenced by the following (all the other factors are constant):

(a) the standard deviation s. (b) the sample size N. (c) the confidence level.

13. The content of Cr in high-alloy steels was investigated by potentiometric titration. Following

are their results (%*w/w* Cr) for the analysis of a single reference steel: 16.968, 16.922, 16.840, 16.883, 16.887, 16.977, 16.857, 16.728.

Calculate the 95% confidence interval about the mean.

14. The following replicate calcium determinations on a blood sample using atomic absorption spectrophotometry (AAS) and a new colorimetric method were reported. Is there a significant difference in the precision of the two methods at the 95% confidence level?

AAS (mg/dl): 10.9, 10.1, 10.6, 11.2, 10.0

Colorimetric (mg/dl): 10.5, 9.7, 11.5, 11.6, 9.3, 10.1, 11.1

15. To test a spectrophotometer for its accuracy, a solution of 0.06006mg/L $K_2Cr_2O_7$ in 5.0mmol/L H_2SO_4 is prepared and analyzed. This solution has a known absorbance of 0.640 at 350.0nm in a 1.0cm cell when using 5.0mmol/L H_2SO_4 as a reagent blank. Several aliquots of the solution are analyzed with the following results:

0.639, 0.638, 0.640, 0.639, 0.640, 0.639, 0.638

Determine whether there is a significant difference between the experimental mean and the expected value at the 99% confidence level.

16. Apply the Q test to the following data set to determine whether the outlying result 85.10 should be retained or rejected at the 95% confidence level: 85.10, 84.62, 84.65, 84.70.

17. Determination of phosphorous in blood serum gave results of 4.40, 4.42, 4.60, 4.48, and 4.50×10^{-3}mg/L. Apply the G test to determine whether the 4.60×10^{-3}mg/L result should be retained or rejected at the 95% confidence level.

Chapter 3 Gravimetric Analysis

学习目标

知识要求

1. 掌握 沉淀重量法对沉淀形式和称量形式的要求，晶形沉淀和非晶形沉淀的沉淀条件；换算因数和质量百分数的计算。

2. 熟悉 影响沉淀溶解度的主要因素；造成沉淀不纯的因素及减免措施。

3. 了解 不同类型沉淀的特点与沉淀的形成过程；挥发重量法与萃取重量法的原理与应用。

能力要求

通过学习重量分析法原理、方法、特点和分类等知识，具备应用重量法在药物质量标准纯度检查及药物含量测定的能力。

PPT

Section 1 Introduction to Gravimetric Analysis

Gravimetric analysis (重量分析法) is an analytical method that uses only measurements of mass and information on reaction stoichiometry to determine the amount of an analyte in a sample. Gravimetric analysis is one of the most accurate and precise methods of macroquantitative analysis. In this process the analyte is selectively converted to an insoluble form. The separated precipitate is dried or ignited, possibly to another form, and is accurately weighed. From the weight of the precipitate and a knowledge of its chemical composition, we can calculate the weight of analyte in the desired form.

A significant advantage of gravimetric analysis is that determining the mass of a substance is one of the most accurate measurements that can be made. Gravimetric analysis does not require a series of standards for calculation of an unknown since calculations are based only on atomic or molecular weights. Only a precise analytical balance is needed for measurements. Many of the gravimetric methods that we discuss in this chapter can be carried out with errors of less than 0.2%. Traditional gravimetric analysis is also inexpensive to conduct and requires only a minimal amount of equipment, such as a high-quality laboratory balance and perhaps a drying oven. While gravimetric analysis is tedious and time consuming, not applicable to the determination of trace and trace components and the control analysis of production, it has already been gradually replaced by other fast and sensitive methods. At present, gravimetric analysis, been included in the pharmacopeia as a legal method of determination, is still used

in the analysis and examination of drugs, for example, determination of the content of components, drying loss, burning residue, and Chinese herbal ash. Gravimetric analysis, due to its high degree of accuracy, can also be used to calibrate other instruments in place of reference standards. Gravimetric analysis is used mainly in modern laboratories when high accuracy is absolutely essential and time is of little concern. For instance, the National Institute of Standards and Technology uses gravimetric analysis as a "gold standard technique", or a reference method, to evaluate the accuracy of other analytical techniques. Thus, Gravimetric analysis is still indispensable in analytical chemistry.

Gravimetric analysis consists of two processes: separation and weighing, in which separation is the most crucial step. Different separation methods are adopted. According to the separation method, the weight method is divided into volatilization gravimetry, extraction gravimetry, precipitation gravimetry, and electrogravimetry. The first three methods are mainly applied in drug inspection.

PPT

Section 2 Volatilization Gravimetry

In volatilization gravimetry, thermal or chemical energy is used to decompose the sample containing the analyte. The volatile products of the decomposition reaction may be trapped to a constant weight and weighed to provide quantitative informations. Alternatively, the residue remaining when decomposition is complete may be weighed to a constant weight. Constant weight (恒重) is defined as a weight difference of less than 0.3mg after two consecutive drying or burning of the sample.

1. Direct analyses

Direct analyses is based on the volatile products of the decomposition reaction may be trapped to a constant weight and weighed to provide quantitative information. The most important application of volatilization gravimetry to the direct analysis of organic materials is an elemental analysis. When burned in a stream of pure O_2, many elements, such as carbon and hydrogen, are released as gaseous combustion products, such as CO_2 and H_2O. The combustion products are passed through preweighed tubes containing appropriate absorbents. The increase in the mass of these tubes provides a direct indication of the mass percent of carbon and hydrogen in the organic material. As a final example, the determination of carbon in steels and other metal alloys can be determined by heating the sample. The carbon is converted to CO_2, which is collected in an appropriate absorbent trap, providing a direct measure of the amount of C in the original sample.

2. Indirect analyses

Indirect analyses is based on the residue remaining when decomposition is complete may be weighed until a constant weight is obtained.

Indirect analyses based on the weight of the residue remaining after volatilization are commonly used in determining moisture in drugs according to the drug purity testing items of pharmacopeia.

Moisture is determined by drying a preweighed sample with an infrared lamp or in a low-temperature oven (<110°C). The difference between the original weight and the weight after drying equals the mass of water lost. The analyte includes hygroscopic water, crystalline water and volatile substances under such conditions.

Indirect analyses are also used in determining of dissolved solids in water. In this method a sample of the containing water is transferred to a weighed dish and dried to a constant weight at either 103–105°C. Samples dried at the lower temperature retain some occluded water. For example, berberine hydrochloride, which had not been dried, was measured to contain 88.54% of berberine hydrochloride and corresponds to a drying weight loss of 10.12%. Then the content of dried products could be converted as follows:

$$\frac{88.54}{100-10.12} \times 100\% = 98.51\%$$

In the analysis of traditional Chinese medicine, ash content is one of the inspection items of Chinese medicinal materials quality control. The sample is generally weighed, placed in an appropriate crucible, and the organic material is carefully removed by combustion. The crucible containing the residue is then heated to a constant weight using either a burner or an oven. According to the pharmacopeia, the obtained ash is treated with sulfuric acid before burning, so that their composition is transformed into a more stable form of oxide and sulfate, known as the burning residue.

PPT

Section 3 Extraction Gravimetry

Extraction gravimetric is a method to separate the analyte from its matrix with a suitable solvent according to their different distribution ratios in two immiscible solvents. After extraction, the solvent can be evaporated and the mass of the extracted analyte can be determined.

In a simple liquid-liquid extraction, the solute is partitioned between two immiscible phases. In most cases one of the phases is aqueous, and the other phase is an organic solvent such as diethyl ether or chloroform. Because the phases are immiscible, they form two layers, with the denser phase on the bottom. The solute is initially present in one phase, but after extraction it is present in both phases. The efficiency of a liquid-liquid extraction is determined by the equilibrium constant for the solute's partitioning between the two phases.

When a phase containing a solute, *S*, is brought into contact with a second phase, the solute partitions itself between the two phases.

$$S_{\text{phase 1}} \rightleftharpoons S_{\text{phase 2}}$$

The equilibrium constant for above reaction

$$D = \frac{[S_{\text{phase 1}}]}{[S_{\text{phase 2}}]} \tag{3-1}$$

D is called the distribution constant, or partition coefficient. If *D* is sufficiently large, then the solute will move from phase 2 to phase 1. The solute will remain in phase 1, however, if the partition coefficient is sufficiently small. If a phase containing two solutes is brought into contact with a second phase, and

D is favorable for only one of the solutes, then a separation of the solutes may be possible. The physical states of the two phases are identified when describing the separation process, with the phase containing the sample listed first.

According to the properties of free alkaloids and organic acids soluble in organic solvents, but their salt soluble in water and insoluble in organic solvents, the alkaloids and organic acids in Chinese medicinal materials could be separated from their matrix by adjusting their chemical forms in different pH of extracted solution, and then determine their content by extraction gravimetric. For example, determination of total alkaloids in belladonna: a certain amount of belladonna powder is added to the lime water and extracted by benzene reflux, the obtained rough extract is then extracted again using 0.5% sulfuric acid by liquid-liquid extraction, in which sodium hydroxide was added in acid water layer to adjust the pH in the range of 11–11.5, and benzene was used to fractional extraction until the alkaloid was extracted completely. All of the benzene extracts were combined and filtration. The total alkaloid content in belladonna can be calculated by evaporation, drying and weighing of filtrates.

Section 4 Precipitation Gravimetry

1. Process and characteristics of precipitation gravimetry

Precipitation gravimetry is based on the formation of an insoluble compound following the addition of a precipitating reagent, or precipitant, to a solution of the analyte. The insoluble compounds were separated from the solution in precipitate formed (沉淀形式), and then the precipitation was filtered, washed free of impurities, converted to precipitate weighed (称量形式) of known composition by suitable heat treatment, and weighed to calculate the percentage content of the measured components. The process of precipitation method as follows in Figure 3-1:

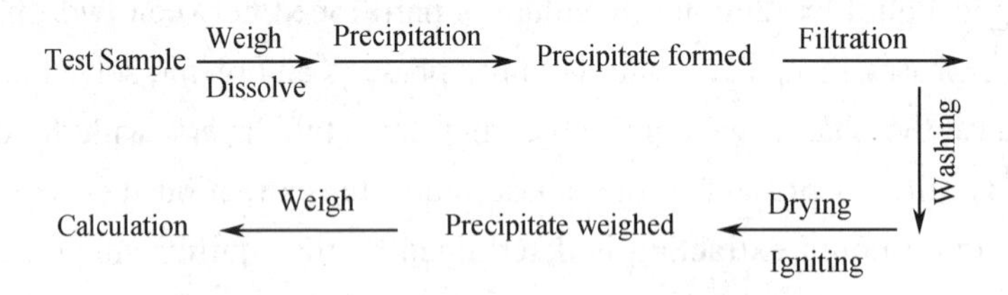

Figure 3-1 The process of precipitation method

2. Preparation of precipitation

The ideal product of a gravimetric analysis should be pure, insoluble, and easily filterable, and should possess a known composition. Few substances meet these requirements, but appropriate techniques can help to optimize properties of gravimetric precipitates.

2.1 Selection of precipitant

The appropriate choose of gravimetric precipitating agent is benefit to obtain an ideal product of a

gravimetric analysis.

(1) Ideally, a gravimetric precipitating agent should react specifically or at least selectively with the analyte. Specific reagents, which are rare, react only with a single chemical species. Selective reagents, which are more common, react with a limited number of species.

(2) The ideal precipitant would react with the analyte to give a product that is of sufficiently low solubility that no significant loss of the analyte occurs during filtration and washing.

(3) Choose volatile precipitant as far as possible to remove excess precipitants in drying or burning to make sure the precipitation is pure;

(4) Better use an organic precipitant, which could react with metal ions to form organic insoluble organic salts or insoluble chelates. Compared with the inorganic precipitating agent, the advantages of the organic precipitating agent are as follows: ①Some of these are very selective. An example of a selective reagent is $AgNO_3$. The only common ions that it precipitates from acidic solution are Cl^-, Br^-, I^-, and SCN^-. Dimethylglyoxime is a specific reagent that precipitates only Ni^{2+} from alkaline solutions. ②Giving precipitates with very low solubility in water. ③The resulting precipitates have a favorable gravimetric factor resulting in large crystalline particle size, less adsorbed to inorganic impurities, easy to filter and wash free of impurities. ④The precipitate has a large molar mass, which is beneficial to reduce the relative error of weighing and improve the accuracy of analysis. ⑤The composition of precipitation is constant and can be weighed after drying without high temperature burning, which simplifies the analysis steps.

2.2 Requirements for precipitation by precipitation gravimetric

2.2.1 The precipitate formed and precipitate weighed

In precipitation gravimetric, the precipitate formed is the chemical composition of precipitation, the precipitate weighed is the chemical composition of the precipitation underwent by suitable heat treatment. The precipitate formed is not always the same as the precipitate weighed.

$$SO_4^{2-} + BaCl_2 \rightarrow BaSO_4\downarrow \xrightarrow[\text{800°C igniting the precipitate}]{\text{Filtering, Washing, Drying}} BaSO_4$$

$$Mg^{2+} + (NH_4)_2HPO_4 \rightarrow MgNH_4PO_4\downarrow \xrightarrow[\text{1100°C igniting the precipitate}]{\text{Filtering, Washing, Drying}} Mg_2P_2O_7$$

In the above examples, both of the precipitate formed and the precipitate weighed of the former reaction are $BaSO_4$; the precipitate formed of the latter reaction is $MgNH_4PO_4$, and different from the precipitate weighed of $Mg_2P_2O_7$.

2.2.2 Requirements for precipitate formed by precipitation method

(1) The precipitate formed must be of low solubility so that the solution loss of the analyte should be lower than the weighting error (± 0.2mg) of analytical balance;

(2) The precipitate formed must be high purity free of contaminations;

(3) The precipitate formed must be in a form that is easy to separate from the reaction mixture;

(4) The precipitate formed must be easy to convert to the precipitate weighed.

2.2.3 Requirements for precipitate weighed by precipitation method

(1) The precipitate weighed must be of known composition if its mass is to accurately reflect the analyte's mass;

(2) The precipitate weighed must be stable, unreactive with constituents of the atmosphere;

(3) The precipitate weighed should have enough high molar weight, which benefit to increase the

mass of the precipitate weighed, reduce the weighing error, and improve the sensitivity and accuracy of the analysis.

2.3 The solubility of precipitation and its influencing factors

An accurate precipitation gravimetric method requires that the precipitate's solubility be minimal. Many total analysis techniques can routinely be performed with an accuracy of better than ±0.1%. To obtain this level of accuracy, the isolated precipitate must account for at least 99.9% of the analyte. By extending this requirement to 99.99% we ensure that accuracy is not limited by the precipitate's solubility. Solubility losses are minimized by carefully controlling the composition of the solution in which the precipitate forms. This, in turn, requires an understanding of the relevant equilibrium reactions affecting the precipitate's solubility.

2.3.1 The effect of a common ion on the solubility of a precipitate

The solubility of an ionic precipitate decreases, when a soluble compound containing one of the ions of the precipitate is added to the solution. This behavior is called the common ion effect (同离子效应), which is a mass-action effect predicted from Le Châtelier's principle (勒夏特列原理，又称为化学平衡移动原理) and is demonstrated by the effect of excess barium ion on the solubility of $BaSO_4$ as shown in Figure 3-2.

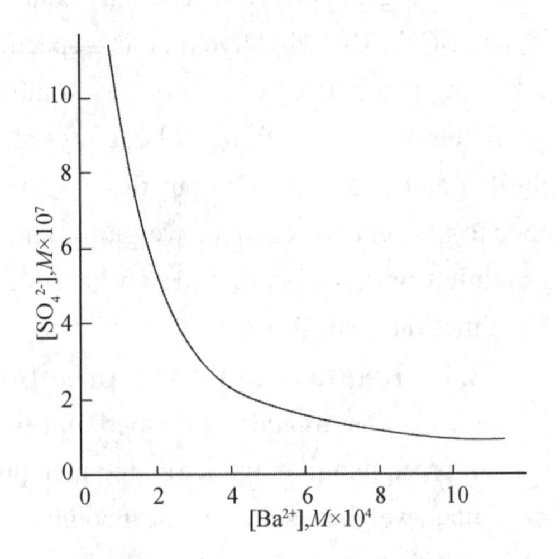

Figure 3-2 Predicted effect of excess barium ion on solubility of $BaSO_4$. Sulfate concentration is amount in equilibrium and is equal to $BaSO_4$ solubility. In absence of excess barium ion, solubility is 10^{-5}mol/L

We take advantage of the common ion effect to decrease the solubility of a precipitate in gravimetric analysis. Under normal circumstances, the precipitant should be excessive 50%–100%; In the case of the precipitant which is not volatile, the precipitant is generally appropriate to excess 20%–30%. However, excess precipitant may cause side reactions such as diverse ion effect, acid effect, and coordination effect, which will increase the solubility.

2.3.2 The effect of a diverse ion (异离子效应) on the solubility of a precipitate

The presence of diverse salts will generally increase the solubility of precipitates due to the shielding (or decrease in the activity) of the dissociated ion species. The diverse ion effect is greater with the increased ions strengths of the extraneous strong electrolyte. For the given electrolyte, the ionic strength will be proportional to the concentration. Strong acids that are completely ionized are treated in the same manner as salts. If the acids are partially ionized, then the concentration of the ionized species must be estimated from the ionization constant before the ionic strength is computed. Very weak acids can usually be considered to be nonionized and do not contribute to the ionic strength.

In precipitation gravimetric analysis, as the precipitant is usually a strong electrolyte, the effect of diverse ion effect should also be considered while using the common ion effect to ensure the complete precipitation. When the precipitator is appropriately excessive, the common ion effect plays a leading role, and the solubility of the precipitation decreases with the increase of the concentration of precipitant. When the concentration of precipitant in the solution reaches a certain value, the solubility of the

precipitation reaches the minimum value. If the precipitator continues to increase, the solubility of the precipitation will increase due to the effect of the diverse ion. Therefore, an excessive amount of precipitant should be appropriate.

For example, Pb^{2+} can be determined gravimetrically by adding Na_2SO_4 as precipitant, forming a precipitate of $PbSO_4$.

$$Pb^{2+} + SO_4^{2-} \rightleftharpoons PbSO_4 \downarrow$$

If this is the only reaction considered, we would falsely conclude that the precipitate's solubility, S_{PbSO_4}, is given by

$$S_{PbSO_4} = [Pb^{2+}] = \frac{K_{sp}}{[SO_4^{2-}]}$$

and that solubility losses may be minimized by adding a large excess of SO_4^{2-}. In fact, as shown in Table 3-1, when the concentration of Na_2SO_4 is less than or equal to 0.04mol/L, the common ion effect plays a leading role, the solubility of $PbSO_4$ decreases with the increase of the concentration of Na_2SO_4. Adding a high concentration of Na_2SO_4 ($c_{Na_2SO_4}$ > 0.04mol/L) eventually increases the precipitate's solubility due to the diverse ion effect.

Table 3-1 The solubility of $PbSO_4$ in the different concentrations of Na_2SO_4

Na_2SO_4 (mol/L)	0	0.001	0.01	0.02	0.04	0.100	0.200
$PbSO_4$ (mol/L)	0.15	0.024	0.016	0.014	0.013	0.016	0.023

It should be pointed out that if the solubility of precipitation is very small, the effect of diverse ion effect is generally small and can be ignored. Diverse ions effects are considered only when the solubility of the precipitate is large and the ionic strength of the solution is high.

2.3.3 The effect of acid (酸效应) on the solubility of a precipitate

Acids frequently affect the solubility of a precipitate. For example, hydroxide precipitates, such as $Fe(OH)_3$, are more soluble at lower pH levels at which the concentration of OH^- is small. The effect of pH on solubility is not limited to hydroxide precipitates, but also affects precipitates containing basic or acidic ions. The solubility of CaC_2O_4 is pH-dependent because $H_2C_2O_4$ is a weak base. As the H^+ concentration increases, it competes more effectively with the metal ion of interest for the precipitating agent. With less free reagent available, and a constant K_{sp}, the solubility of the salt must increase:

$$Ca^{2+} + C_2O_4^{2-} \rightleftharpoons CaC_2O_4 \quad \text{(desired reaction)}$$

$$H^+ + C_2O_4^{2-} \rightleftharpoons HC_2O_4^- \quad \text{(competing reaction)}$$

$$H^+ + HC_2O_4^- \rightleftharpoons H_2C_2O_4 \quad \text{(competing reaction)}$$

$$CaC_2O_4 + H^+ \rightleftharpoons Ca^{2+} + HC_2O_4^- \quad \text{(overall reaction)}$$

$$CaC_2O_4 + 2H^+ \rightleftharpoons Ca^{2+} + H_2C_2O_4 \quad \text{(overall reaction)}$$

As the acidity of the solution increases and the equilibrium moves in the direction of formation $HC_2O_4^-$ and $H_2C_2O_4$, the solubility of CaC_2O_4 is increasing. The effect of acidity on precipitation solubility is complex. The acidity of the solution has an important effect on the solubility of the insoluble salts of polyacid and weak acid. But, it has little effect on the solubility of the insoluble salts of strong acid.

2.3.4 The effect of complexation (配位效应) on the solubility of a precipitate

The solubility of a precipitate can be improved by adding a ligand capable of forming a soluble complex with one of the precipitate's ions. For example, the solubility of AgCl increases in the presence of NH_3 due to the formation of the soluble $Ag(NH_3)_2^+$ complex.

$$AgCl + 2NH_3 \rightleftharpoons Ag(NH_3)_2^+ + Cl^-$$

The precipitation of AgCl was obtained from chloride solution reacts with Ag^+ solution. When the concentration of chloride solution is appropriately excessive, the common ion effect plays a major role and the solubility of AgCl is reduced. However, if the precipitator continues to increase, the solubility of the precipitation will increase due to the formation of the soluble $Ag(NH_3)_2^+$ complex.

2.3.5 The effect of other factors on the solubility of a precipitate

Solubility can often be decreased by using a nonaqueous solvent. A precipitate's solubility is generally greater in aqueous solutions because of the ability of water molecules to stabilize ions through solvation. The poorer solvating ability of nonaqueous solvents, even those that are polar, leads to a smaller solubility product. For example, $PbSO_4$ has a K_{sp} of 2.53×10^{-8} in H_2O, whereas in a 50 : 50 mixture of H_2O/ethanol the K_{sp} at 2.6×10^{-12} is four orders of magnitude smaller.

2.4 The purity of precipitation and its influencing factors

Precipitation gravimetry is based on a known stoichiometry between the analyte's mass and the mass of a precipitate. It follows, therefore, that the precipitate must be free from impurities. Since precipitation typically occurs in a solution rich in dissolved solids, the initial precipitate is often impure. Any impurities present in the precipitate's matrix must be removed before obtaining its weight.

2.4.1 Coprecipitation (共沉淀)

Precipitates tend to carry down from the solution other constituents that are normally soluble, causing the precipitate to become contaminated. This process is called coprecipitation. The process may be equilibrium based or kinetically controlled. There are a number of ways in which a foreign material may be coprecipitated.

(1) Surface Adsorption (表面吸附). The surface of the precipitate will have a primary adsorbed layer of the lattice ions in excess. This results in surface adsorption, the most common form of the precipitate contamination. For example, after $BaSO_4$ is completely precipitated in a solution containing excess Ba^{2+}, H^+ and NO_3^-, the surface of the particle has excess positive charge due to the adsorption of extra barium ions on exposed the sulfate anions, and this will form the primary layer that is strongly adsorbed and is integral part of the crystal. The counter ion will be a foreign anion, say, a nitrate anion, two for each barium. The net effect then is an adsorbed layer of barium nitrate, an equilibrium-based process. These adsorbed layers can often be removed by washing, or they can be replaced by ions that are readily volatilized. Digestion reduces the surface area and the amount of adsorption.

(2) Occlusion and Inclusion (吸留与包藏). In the process of occlusion, material that is not part of the crystal structure is trapped within a crystal. For example, water may be trapped in pockets when AgCl crystals are formed, and this can be removed to a degree by dissolution and recrystallization. If such mechanical trapping occurs during a precipitation process, the water will contain dissolved impurities. Inclusion occurs when ions, generally of similar size and charge, are trapped within the crystal lattice (isomorphous inclusion, as with K^+ in NH_4MgPO_4 precipitation). These are not equilibrium processes. Occluded or included impurities are difficult to remove. Digestion may help some but is not completely effective. The impurities cannot be removed by washing. Purification by dissolving and reprecipitating is helpful.

(3) Isomorphous Replacement (生成混晶). Two compounds are said to be isomorphous if they have the same type of formula and crystallize in similar geometric forms. When their lattice dimensions

are about the same, one ion can replace another in a crystal, resulting in a mixed crystal. This process is called isomorphous replacement or isomorphous substitution. For example, in the precipitation of Mg^{2+} as magnesium ammonium phosphate, K^+ has nearly the same ionic size as NH_4^+ and can replace it to form magnesium potassium phosphate. Isomorphous replacement, when it occurs, causes major interference, and little can be done about it. Precipitates in which it occurs are seldom used analytically. Chloride cannot be selectively determined by precipitation as AgCl, for example, in the presence of other halides and vice versa. Mixed crystal formation is a form of equilibrium precipitate formation, although it may be influenced by the rate of precipitation. Such a mixed precipitate is akin to a solid solution. The mixed crystal may be spatially homogeneous if the crystal is in equilibrium with the final solution composition (homogeneous coprecipitation) or heterogeneous if it is in instaneous equilibrium with the solution as it forms (heterogeneous coprecipitation), as the solution composition changes during precipitation.

2.4.2 Postprecipitation (后沉淀)

Sometimes, when the precipitate is allowed to stand in contact with the mother liquor, a second substance will slowly form a new precipitate with the precipitating reagent. This is called postprecipitation. For example, when calcium oxalate is precipitated in the presence of magnesium ions, magnesium oxalate does not immediately precipitate because it tends to form supersaturated solutions. But it will precipitate if the solution is allowed to stand too long before being filtered. Similarly, copper sulfide will precipitate in acid solution in the presence of zinc ions without zinc sulfide being precipitated, but eventually zinc sulfide will precipitate. Postprecipitation is a slow equilibrium process. It means the amount of new precipitate will increased with the longer time of the precipitate stand in the mother liquor. Therefore, the time of the precipitate stand in the mother liquor must be shortened to eliminate the postprecipitation.

2.5 Formation and conditions of precipitation

2.5.1 Precipitation size

The size of the precipitate's particles determines the ease and success of filtration. Smaller, colloidal particles are difficult to filter because they may readily pass through the pores of the filtering device. Large, crystalline particles, however, are easily filtered. According to their different physical properties, precipitation can be roughly divided into crystalline precipitation (Diameter: 0.1–1μm) and amorphous precipitation (amorphous precipitation) (<0.02μm). Precipitated particles should not be so small that they clog or pass through the filter. Larger crystals have less surface area to which impurities can become attached. At the other extreme is a colloidal suspension of particles that have diameters in the approximate range 1-100nm and pass through most filters. By carefully controlling the precipitation reaction we can significantly increase a precipitate's average particle size.

2.5.2 Formation of precipitation

Crystallization occurs in two phases: nucleation and particle growth.

(1) Nucleation: in nucleation, solutes are thought to form a disorganized cluster of sufficient size, which then reorganizes into an ordered structure capable of growing into larger particles. Nucleation can occur on suspended impurity particles or scratches on a glass surface. When Fe^{3+} reacts with 0.1mol/L tetramethylammonium hydroxide at 25°C, nuclei of hydrated $Fe(OH)_3$ are 4nm in diameter and contain~50 Fe atoms.

(2) Particle growth: in particle growth, molecules or ions condense onto the nucleus to form a larger crystal. $Fe(OH)_3$ nuclei grow into plates with lateral dimensions of ~30nm×7nm after 15 min at 60°C.

Larger particles form when the rate of particle growth exceeds the rate of nucleation. A solute's relative supersaturation (相对过饱和度, RSS), can be expressed as

$$\mathrm{RSS} = \frac{Q-S}{S} \tag{3-2}$$

Where Q is the solute's actual concentration, S is the solute's expected concentration at equilibrium, and $Q - S$ is a measure of the solute's supersaturation when precipitation begins. This ratio, $(Q-S)/S$, relative supersaturation, is also called the von Weimarn ratio. A large, positive value of RSS indicates that a solution is highly supersaturated. Such solutions are unstable and show high rates of nucleation, producing a precipitate consisting of numerous small particles. When RSS is small, precipitation is more likely to occur by particle growth than by nucleation.

High relative supersaturation → many small crystals (high surface area)

Low relative supersaturation → fewer, larger crystals (low surface area)

Equation 3-2 shows that we can minimize RSS by either decreasing the solute's concentration or increasing the precipitate's solubility. A precipitate's solubility usually increases at higher temperatures, and adjusting pH may affect a precipitate's solubility if it contains an acidic or basic anion. Temperature and pH, therefore, are useful ways to increase the value of S. Conducting the precipitation in a dilute solution of analyte, or adding the precipitant slowly and with vigorous stirring are ways to decrease the value of Q.

There are, however, practical limitations to minimizing RSS. Precipitates that are extremely insoluble, such as $Fe(OH)_3$ and PbS, have such small solubilities that a large RSS cannot be avoided. Such solutes inevitably form small particles. In addition, conditions that yield a small RSS may lead to a relatively stable supersaturated solution that requires a long time to fully precipitate. For example, almost a month is required to form a visible precipitate of $BaSO_4$ under conditions in which the initial RSS is 5.

An increase in the time required to form a visible precipitate under conditions of low RSS is a consequence of both a slow rate of nucleation and a steady decrease in RSS as the precipitate forms. One solution to the latter problem is to chemically generate the precipitant in solution as the product of a slow chemical reaction. This maintains the RSS at an effectively constant level. The precipitate initially forms under conditions of low RSS, leading to the nucleation of a limited number of particles. As additional precipitant is created, nucleation is eventually superseded by particle growth. This process is called homogeneous precipitation.

Two general methods are used for homogeneous precipitation. If the precipitate's solubility is pH-dependent, then the analyte and precipitant can be mixed under conditions in which precipitation does not occur. The pH is then raised or lowered as needed by chemically generating OH^- or H_3O^+. For example, the hydrolysis of urea can be used as a source of OH^-.

$$CO(NH_2)_2 + H_2O \xrightleftharpoons{90\text{–}100℃} CO_2 + 2NH_3$$

$$NH_3 + H_2O \rightleftharpoons NH_4^+ + OH^-$$

The hydrolysis of urea is strongly temperature-dependent, with the rate being negligible at room temperature. The rate of hydrolysis, and thus the rate of precipitate formation, can be controlled by adjusting the solution's temperature. Precipitates of $BaCrO_4$, for example, have been produced in this manner. In the second method of homogeneous precipitation, the precipitant itself is generated by a chemical reaction. For example, Ba^{2+} can be homogeneously precipitated as $BaSO_4$ by hydrolyzing sulphamic acid to produce SO_4^{2-}.

$$NH_2SO_3H + 2H_2O\,(沸水) \rightleftharpoons NH_4^+ + H_3O^+ + SO_4^{2-}$$

Homogeneous precipitation affords the dual advantages of producing large particles of precipitate that are relatively free from impurities. These advantages, however, may be offset by increasing the time needed to produce the precipitate, and a tendency for the precipitate to deposit as a thin film on the container's walls. The latter problem is particularly severe for hydroxide precipitates generated using urea.

An additional method for increasing particle size deserves mention. When a precipitate's particles are electrically neutral, they tend to coagulate into larger particles. Surface adsorption of excess lattice ions, however, provides the precipitate's particles with a net positive or negative surface charge. Electrostatic repulsion between the particles prevents them from coagulating into larger particles.

Consider, for instance, the precipitation of AgCl from a solution of $AgNO_3$, using NaCl as a precipitant. Early in the precipitation, when NaCl is the limiting reagent, excess Ag^+ ions chemically adsorb to the AgCl particles, forming a positively charged primary adsorption layer (Figure 3-3). Anions in solution, in this case NO_3^- and OH^-, are attracted toward the surface, forming a negatively charged secondary adsorption layer that balances the surface's positive charge. The solution outside the secondary adsorption layer remains electrically neutral. Coagulation cannot occur if the secondary adsorption layer is too thick because the individual particles of AgCl are unable to approach one another closely enough.

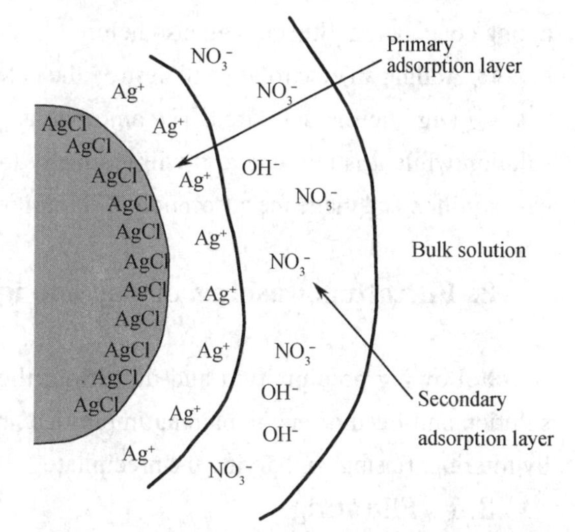

Figure 3-3 Schematic model of the solid-solution interface at a particle of AgCl in a solution containing excess $AgNO_3$

Coagulation can be induced in two ways: by increasing the concentration of the ions responsible for the secondary adsorption layer or by heating the solution. One way to induce coagulation is to add an inert electrolyte, which increases the concentration of ions in the secondary adsorption layer. With more ions available, the thickness of the secondary absorption layer decreases. Particles of precipitate may now approach one another more closely, allowing the precipitate to coagulate. Heating the solution and precipitate provides a second way to induce coagulation. As the temperature increases, the number of ions in the primary adsorption layer decreases, lowering the precipitate's surface charge. In addition, increasing the particle's kinetic energy may be sufficient to overcome the electrostatic repulsion preventing coagulation at lower temperatures.

2.5.3 Conditions of precipitation

(1) Precipitation conditions for amorphous precipitation. Obviously, keeping *RSS* low and *S* high during precipitation is conducive to obtain larger crystals as shown in Equation 3-2. So, techniques that promote crystalline particle growth include:

① Raising the temperature to increase solubility and thereby decrease supersaturation;

② Adding precipitant slowly with vigorous mixing, to prevent a local, highly supersaturated condition where the stream of precipitant first enters the analyte;

③ Using a large volume of solution so that concentrations of analyte and precipitant are low;

④ Maintaining the precipitate in equilibrium with its supernatant solution for an extended time. This process is called digestion and may be carried out at room temperature or elevated temperature. During digestion, the dynamic nature of the solubility–precipitation equilibrium, in which the precipitate dissolves and re-forms, ensures that occluded material is eventually exposed to the supernatant solution. Since the rate of dissolution and reprecipitation is slow, the chance of forming new occlusions is minimal.

(2) Precipitation conditions for amorphous precipitation. Many precipitates do not give a favorable *RSS*, especially very insoluble ones. Hence, it is impossible to yield a crystalline precipitate (a small number of large particles), and the precipitate is first colloidal (a large number of small particles). Therefore, the conditions for amorphous precipitation are mainly trying to destroy the colloid, prevent gelatinization, accelerate coagulation.

① Increasing the concentration of the precipitant and heating the solution could induce coagulation;

② Heating tends to decrease adsorption and the effective prevent the formation of colloid, thereby aiding coagulation. Stirring will also help;

③ Adding an electrolyte will destroy the colloid and promote coagulation;

④ Digestion is undesired. The amorphous precipitation should be separated immediately from the solution while it is hot, due to it will gradually lose water in solution so that it is more cohesive and not easy to filter, and make the adsorption of impurities difficult to wash off.

3. Filtrating, washing, drying and igniting of precipitation

Following precipitation and digestion, the precipitate must be separated from the supernatant solution and freed of any remaining impurities, including residual solvent. These tasks are accomplished by filtering, rinsing, and drying the precipitate.

3.1 Filtrating

The precipitate is separated from the solution by filtration using either filter paper (滤纸) or a filtering crucible. The most common filtering medium is cellulose-based filter paper, which is classified according to its filtering speed, its size, and its ash content on ignition. Filtering speed is a function of the paper's pore size, which determines the particle sizes retained by the filter. Filter paper is rated as fast (retains particles > 20–25mm), medium fast (retains particles > 16mm), medium (retains particles > 8mm), and slow (retains particles > 2–3mm). The proper choice of filtering speed is important. If the filtering speed is too fast, the precipitate may pass through the filter paper resulting in a loss of precipitate. On the other hand, the filter paper can become clogged when using a filter paper that is too slow.

Filter paper is hygroscopic and is not easily dried to a constant weight. As a result, in a quantitative procedure the filter paper must be removed before weighing the precipitate. This is accomplished by carefully igniting the filter paper. Following ignition, a residue of noncombustible inorganic ash remains that contributes a positive determinate error to the precipitate's final mass. For quantitative analytical procedures a low-ash filter paper must be used.

An alternative method for filtering the precipitate is a filtering crucible. The most common is a fritted glass crucible containing a porous glass disk filter. Fritted glass crucibles are classified by their porosity: coarse (retaining particles > 40–60mm), medium (retaining particles > 10–15mm), and fine (retaining particles > 4–5.5mm). Another type of filtering crucible is the Gooch crucible, a porcelain crucible with a perforated bottom. A glass fiber mat is placed in the crucible to retain the precipitate, which is transferred to the crucible in the same manner described for filter paper. The supernatant is drawn through the crucible with the assistance of suction from a vacuum aspirator or pump.

3.2 Washing

Filtering removes most of the supernatant solution. Residual traces of the supernatant, however, must be removed to avoid a source of determinate error. Washing the precipitate to remove this residual material must be done carefully to avoid significant losses of the precipitate. Of greatest concern is the potential for solubility losses. Usually the washing medium is selected to ensure that solubility losses are negligible. In many cases this simply involves the use of cold solvents or rinse solutions containing organic solvents such as ethanol. Precipitates containing acidic or basic ions may experience solubility losses if the rinse solution's pH is not appropriately adjusted. When coagulation plays an important role in determining particle size, a volatile inert electrolyte is often added to the rinse water to prevent the precipitate from reverting into smaller particles that may not be retained by the filtering device. This process of reverting to smaller particles is called peptization. The volatile electrolyte is removed when drying the precipitate. When rinsing a precipitate there is a trade-off between introducing positive determinate errors due to ionic impurities from the precipitating solution and introducing negative determinate errors from solubility losses. In general, solubility losses are minimized by using several small portions of the rinse solution instead of a single large volume. Testing the used rinse solution for the presence of impurities is another way to ensure that the precipitate is not over-washed. This can be done by testing for the presence of a targeted solution ion and rinsing until the ion is no longer detected in a freshly collected sample of the rinse solution. For example, when Cl^- is known to be a residual impurity, its presence can be tested by adding a small amount of $AgNO_3$ to the collected rinse solution. A white precipitate of AgCl indicates that Cl^- is present and additional rinsing is necessary. Additional washing is not needed, however, if adding $AgNO_3$ does not produce a precipitate.

3.3 Drying and igniting

Finally, after separating the precipitate from its supernatant solution the precipitate is dried to remove any residual traces of rinse solution and any volatile impurities. The temperature and method of drying depend on the method of filtration, and the precipitate's desired chemical form. A temperature of 110°C is usually sufficient when removing water and other easily volatilized impurities. A conventional laboratory oven is sufficient for this purpose. Higher temperatures require the use of a muffle furnace, or a Bunsen or Meker burner, and are necessary when the precipitate must be thermally decomposed before weighing or when using filter paper. To ensure that drying is complete the precipitate is repeatedly dried and weighed until a constant weight is obtained.

The quantitative application of precipitation gravimetry, which is based on conservation of mass, requires that the final precipitate have a well-defined composition. Precipitates containing volatile ions or substantial amounts of hydrated water are usually dried at a temperature that is sufficient to completely remove the volatile species. For example, one standard gravimetric method for the determination of magnesium involves the precipitation of $MgNH_4PO_4 \cdot 6H_2O$. Unfortunately, this precipitate is difficult to dry at lower temperatures without losing an inconsistent amount of hydrated water and ammonia. Instead, the precipitate is dried at temperatures above 1000°C, where it decomposes to magnesium pyrophosphate, $Mg_2P_2O_7$.

4. The calculation of analysis results

The precipitate we weigh is usually in a different form than the analyte whose weight we wish to report. The principles of converting the weight of one substance to that of another are using stoichiometric mole relationships.

If A is analyte, D is of precipitate weighed, the stoichiometric mole relationships can be expressed as follows:

$$\underset{\text{analyte}}{aA} + \underset{\text{precipitant}}{bB} \rightarrow \underset{\text{precipitate formed}}{cC} \xrightarrow{\Delta} \underset{\text{precipitate weighed}}{dD}$$

The stoichiometric mole relationships between A and D can be expressed as follows:

$$n_A = \frac{a}{d} n_D \tag{3-3}$$

When $n=m/M$ is substituted into the above equation, the mass of the analyte is obtained:

$$m_A = \frac{aM_A}{dM_D} m_D \tag{3-4}$$

Where M_A and M_D is the molar mass of analyte A and precipitate weighed D, respectively. We introduced the gravimetric factor (F, 换算因数), which represents the weight of analyte per unit weight of precipitate. It is obtained from the ratio of the formula weight of the analyte to that of the precipitate, multiplied by the moles of analyte per mole of precipitate obtained from each mole of analyte, that is,

$$F = \frac{aM_A}{dM_D} \tag{3-5}$$

$$m_A = F \times m_D \tag{3-6}$$

Where m_A represents the gram of analyte (the desired test substance) and m_D represents the grams of precipitate weighed.

So, if Cl_2 in a sample is converted to chloride and precipitated as AgCl, the weight of Cl_2 that gives 1g of AgCl is ($M_{Cl_2} = 70.90\text{g/mol}$, $M_{AgCl} = 143.32\text{g/mol}$)

$$Cl_2 \rightarrow 2Cl^- \rightarrow 2AgCl\downarrow \xrightarrow{\Delta} 2AgCl\downarrow$$

$$m_{Cl_2} = F \times m_{AgCl} = \frac{aM_{Cl_2}}{dM_{AgCl}} \times m_{AgCl} = \frac{1\times 70.90\text{g/mol}}{2\times 143.32\text{g/mol}} \times 1\text{g} = 0.2473\text{g}$$

Table 3-2 The gravimetric factor (*F*) of selected analyte and precipitate weighed

Analyte	Precipitate formed	Precipitate weighed	*F*
Fe	$Fe(OH)_3 \cdot nH_2O$	Fe_2O_3	$\frac{2M_{Fe}}{M_{Fe_2O_3}}$
MgO	$MgNH_4PO_4$	$Mg_2P_2O_7$	$\frac{2M_{MgO}}{M_{Mg_2P_2O_7}}$
$K_2SO_4 \cdot Al_2(SO_4)_3 \cdot 24H_2O$	$BaSO_4$	$BaSO_4$	$\frac{M_{K_2SO_4 \cdot Al_2(SO_4)_3 \cdot 24H_2O}}{4M_{BaSO_4}}$

In gravimetric analysis, we are generally interested in the percent composition by weight of the analyte in the sample, that is,

$$\%analyte = \frac{\text{weight of analyte}}{\text{weight of sample (g)}} \times 100\% = \frac{m_A}{m_S} \times 100\% = \frac{m_D}{m_S} \times F \times 100\% \quad (3\text{-}7)$$

Where S represents the grams of sample taken for analysis.

【Example 3-1】

A 0.1215g sample of tartaric acid, which was precipitated into calcium tartrate. It was burned into calcium carbonate and then evaporated to dry after excessive HCl treatment. $AgNO_3$ was added to the residue so that the Cl^- was determined in the form of AgCl, and the weight of precipitate weighed AgCl was 0.1103g. Calculate the %*w/w* tartaric acid in the sample.

Solution:

$$H_2C_4H_4O_6 \rightarrow CaC_4H_4O_6 \xrightarrow{\Delta} CaCO_3 \rightarrow CaCl_2 \rightarrow 2AgCl$$

$$H_2C_4H_4O_6 \rightarrow 2AgCl$$

$$\therefore\ H_2C_4H_4O_6\% = \frac{m_{AgCl} \times \dfrac{M_{H_2C_4H_4O_6}}{2 \times M_{AgCl}}}{m_S} \times 100\% = \frac{0.1103 \times \dfrac{150.09}{2 \times 143.32}}{0.1215} \times 100\% = 47.54\%$$

5. The application of precipitation gravimetry

Precipitation gravimetry can be applied to the qualitative analysis of inorganic and organic analytes. Although qualitative applications of precipitation gravimetry have been largely replaced by spectroscopic methods of analysis, they still provides a reliable means for assessing the accuracy of other methods of analysis or for verifying the composition of standard reference materials.

Determination of Na_2SO_4 in mirabilite (The pharmacopeia of People's Republic of China 2015 Edition).

Content determination method: a sample of mirabilite weighing about 0.4g was dissolved in 200ml of water and treated with 1ml of concentrated HCl to boil under stirring. 20ml of hot $BaCl_2$ solution was introduced into the above solution by dripping slowly until the precipitant no longer separate out, then the above solution was heated for 30min in a water bath and stand for 1h. After filtering, washing and drying, the dried precipitates were then ignited to constant weight, and accurately weighed. Gravimetric factor (F) is 0.6086, which represents the weight of sodium sulfate per unit weight of precipitate ($BaSO_4$). So, $m_{Na_2SO_4}$=0.6086×m_{BaSO_4}.

重点小结

重量分析法是通过称量物质的质量确定待测组分含量的方法。是将待测组分分离后，转化为称量形式称量确定含量。重量法具有准确度高，误差小；操作繁琐、费时、灵敏度低，不适宜微量及痕量组分的测定等特点。根据分离方法重量法可分为挥发法、萃取法、沉淀法和电解法。

挥发法是经加热或其他方法使挥发性组分气化逸出或用吸收剂吸收，由样品减失的质量或吸收剂增加的质量计算含量。分为直接挥发法和间接挥发法。常用方法：干燥失重测定 - 通过加热前后样品质量变化测定含水量及挥发性物质；中药灰分测定 - 由高温灼烧后残渣质量测定无机物。药物泡腾制剂中 CO_2 气体释放量也可用挥发法。

沉淀法通过加入沉淀剂，使待测组分以沉淀形式析出，经过滤、洗涤、烘干或灼烧，转化为称量形式称量，计算含量。沉淀法具有重量法的共同特点。沉淀制备的关键是：沉淀作用完全，沉淀纯净，具有易于过滤、洗涤的形态。同离子效应、异离子效应、酸效应、配位效应是影响沉淀溶解度的重要因素。共沉淀、后沉淀是影响沉淀纯度的主要因素。获得良好沉淀形状的条件：晶形沉淀 - 稀、热、搅、慢、陈；非晶形沉淀 - 浓、热、加电解质、不陈化。

沉淀法计算利用换算因数，确定待测组分与称量形式的计量关系：$m_A = F \times m_D$。也可用以解决估计取样量、沉淀剂用量及含量测定结果计算问题。

题库

目标检测

1. Describe the unit operations commonly employed in gravimetric analysis, and briefly indicate the purpose of each?

2. What is digestion of precipitate, and why is it necessary?

3. Outline the optimum conditions for precipitation that will obtain a pure and filterable precipitate.

4. What is coprecipitation? List the different types of coprecipitation, and indicate how they may be minimized or treated.

5. What advantages do organic precipitating agents have?

6. For what pH range will the following precipitates have their lowest solubility?

(a) CaC_2O_4, (b) $PbCrO_4$, (c) $BaSO_4$, (d) $SrCO_3$, (e) ZnS

7. Calculate the gravimetric factors for:

Substance sought	Substance weighed	Gravimetric factors
As_2O_3	Ag_3AsO_4	
$FeSO_4$	Fe_2O_3	
K_2O	$KB(C_6H_5)_4$	
SiO_2	$KAlSi_3O_8$	

8. An ore containing magnetite, Fe_3O_4, was analyzed by dissolving a 1.5419g sample in concentrated HCl, giving a mixture of Fe^{2+} and Fe^{3+}. After adding HNO_3 to oxidize any Fe^{2+} to Fe^{3+}, the resulting solution was diluted with water and the Fe^{3+} precipitated as $Fe(OH)_3$ by adding NH_3. After filtering and washing, the residue was ignited, giving 0.8525g of pure Fe_2O_3. Calculate the %*w/w* Fe_3O_4 in the sample.

9. A 523.1mg sample of impure KBr is treated with excess $AgNO_3$ and 814.5mg AgBr is obtained. What is the purity of the KBr?

Chapter 4 Titrimetric Analysis

学习目标

知识要求

1. **掌握** 滴定分析法基本概念及相关计算。

2. **熟悉** 滴定分析对滴定反应的要求、滴定方式；基准物质及其应具备的条件；标准溶液的配制与标定方法；滴定度的含义及相关计算。

3. **了解** 滴定分析的特点和分类。

能力要求

通过学习标准溶液的相关知识，掌握标准溶液的配制与标定方法；熟练运用滴定分析的相关计算方法。

PPT

Section 1 Introduction to Titrimetric Analysis

Titrimetric analysis (滴定分析法), also called volumetric analysis (容量分析法), is based on determining the volume of a reagent of known concentration that is required to react completely with the analyte. The reagent of exactly known concentration is called the standard solution (标准溶液). The operation process of adding standard solution to analyte is called titration (滴定). Titration is performed by adding exactly the volume of a standard solution needed to react with an unknown quantity of an analyte. The standard solution is also called the titrant (滴定剂); the volume of titrant needed for the titration is carefully measured by means of a buret. From the accurate volume and concentration of the titrant that consumed in the process of the titration, the unknown quantity of an analyte can be calculated.

The stoichiometric point (化学计量点, sp) occurs when the quantity of added titrant is the exact amount necessary for stoichiometric reaction with the analyte. The stoichiometric point, also called the equivalence point, is the theoretical result we seek in a titration. Instead, we can only estimate its position by observing some physical change associated with the condition of chemical equivalence. The position of this change is called the end point of the titration (滴定终点, ep). The end point is determined experimentally by a color change or by a change in an instrumental response. Ideally, the end point and the stoichiometric point coincide. Unfortunately, in most titrations we usually have no obvious indication that the stoichiometric point has been reached. Instead, we stop adding titrant when we reach an end point

of our choosing. The determinate error in a titration due to the difference between the end point and the stoichiometric point is called the titration error (滴定误差, TE) or the end point error (终点误差).

An indicator (指示剂) is a compound with a physical property (usually color) that changes abruptly near the stoichiometric point being used to indicate the end point. Typical indicator changes include the appearance or disappearance of a color, or a change in color. The changes in the relative concentration of analyte or titrant occur in the stoichiometric-point region. Instruments are also used to detect end points. These instruments respond to properties of the solution that change in a characteristic way during the titration. Among such instruments are voltmeters, current meters, conductivity meters, colorimeters, spectrophotometers, and so on.

1. Characteristics and classification of titrimetric analysis

Titrimetric analysis constitutes some of the most important procedures used in quantitative analysis. Titrimetric analysis need only simple equipments, such as buret, but it has high accuracy (commonly, the relative error is below ±0.2%) and usually is used to determine the analyte whose content is higher than 1%. Titration is a rapid, convenient, accurate method and widely used for routine analyses in analytical chemistry. It also plays a important role in the area of drug quality control. For example, titrimetric analysis are the preferred methods to be used in assaying the content of drug substances in Chinese Pharmacopoeia Part II.

According to the various types of titrimetric reactions between the titrant and the analyte, titrimetric analysis can be classified into acid-base titration, precipitation titration, complexometric titration and oxidation-reduction titration.

In this chapter we discuss the general principles and calculations that apply to any titrimetric procedure. The detailed theoretical treatments and actual methods of various types of titration are given in the following chapters.

2. Titration analysis requirements for titrimetric reactions

There are many types of chemical reactions, but not all of them can be used for titrimetric analysis. Chemical reactions used in titrimetric analysis must conform to the following basic requirements.

(1) The reaction should be stoichiometric reacted. All reactions involving the titrant and analyte must be of known stoichiometry, the stoichiometric ratio is assumed to be known exactly without uncertainty. For the usual analytical accuracy, it must be at least 99.9% completed when a stoichiometric amount of titrant has been added. If this is not the case, then the moles of titrant used in reaching the end point cannot tell us how much analyte is in the sample. This is the basis of stoichiometric calculations in titrimetric analysis.

(2) The reaction must occur rapidly. That is, each increment of titrant should be quickly consumed by analyte until the analyte is used up. If the reaction is too slow, heating or catalyst may be used to accelerate the reaction.

(3) A suitable method must be available for determining the end point with an acceptable level of accuracy. The end point can be determined by an indicator color change or by a change in an instrumental response.

3. Different ways of titrimetric analysis

Titrimetric analysis is used in different ways according to the different types of chemical reactions.

3.1 Direct titration

In a direct titration, we add the titrant to analyte until the reaction is complete. A simple example of a titration is an analysis for Ag^+ using thiocyanate, SCN^-, as a titrant. This reaction occurs quickly and is of known stoichiometry. To indicate the titration's end point we add a small amount of Fe^{3+} to the solution containing the analyte. The formation of the red colored $Fe(SCN)^{2+}$ complex signals the end point. This is an example of a direct titration since the titrant reacts with the analyte.

Direct titration is a simple and quick way in titrimetric analysis. Three requirements must need to be met when the chemical reactions used in a direct titrimetric analysis. These are significant limitations and, for this reason, several titration strategies are commonly used.

3.2 Back titration

Back titration is a process to add a known excess of one standard reagent to the analyte and then titrate the excess reagent remaining after its reaction with the analyte with a second standard solution. Back titrations are often required when the reaction is slow, or solid sample, or when a suitable indicator is not available.

For example, the content of solid $CaCO_3$ wants to be determined in an aqueous solution. The acid-base reaction between $CaCO_3$ and HCl is a useful reaction, except that it is too slow for a direct titration. If we add a known amount of a standard solution HCl, such that it is in excess, we can allow the reaction to go to completion. The HCl remaining after reaction with $CaCO_3$ can then be back titrated with a second standard solution NaOH. This type of titration is called a back titration.

$$CaCO_3 + 2HCl\ (excess) = CaCl_2 + H_2O + CO_2\uparrow$$

$$HCl\ (remaining) + NaOH = NaCl + H_2O$$

3.3 Replacement titration

Replacement titration is a titration in which the analyte replaces a substance, and the amount of the replaced substance is determined by a titration with the titrant. In a replacement titration, the titrant and analyte can react, but the reaction is performed with an unknown stoichiometry, such as too much side reactions. The replaced substance can be quantitative replaced from the analyte, and then can be direct titrated with the titrant.

The reaction between potassium dichromate $K_2Cr_2O_7$ and sodium thiosulfate $Na_2S_2O_3$ is performed with an unknown stoichiometry. However, $Na_2S_2O_3$ is a moderately strong reducing agent that has been used to determine oxidizing agents (such as $K_2Cr_2O_7$) by a replacement titration in which iodine I_2 is a replaced substance. Reduction of $K_2Cr_2O_7$ produces a stoichiometrically equivalent amount of I_2. The liberated I_2 is then titrated with a standard solution of $Na_2S_2O_3$ according to the following reactions.

$$Cr_2O_7^{2-} + 6I^- + 14H^+ \rightleftharpoons 3I_2 + 2Cr^{3+} + 7H_2O$$

$$I_2 + 2S_2O_3^{2-} \rightleftharpoons 2I^- + S_4O_6^{2-}$$

3.4 Indirect titration

When a suitable reaction involving the analyte does not exist it may be possible to generate a intermediate that is easily titrated. There is no available direct titration reaction, then an indirect titration may be possible.

The calcium ion Ca^{2+} content can be determined by the following procedure. Precipitate Ca^{2+} with oxalate $C_2O_4^{2-}$ in basic solution. Wash the precipitate with ice-cold water to remove free oxalate and then dissolve the solid in acid to obtain Ca^{2+} and $H_2C_2O_4$ in solution. Heat the solution and titrate $C_2O_4^{2-}$ with standardized potassium permanganate $KMnO_4$. In this sample, several indirect reactions were used to convert Ca^{2+} to $C_2O_4^{2-}$, which we can direct titrate with $KMnO_4$, providing an indirect determination of calcium ion.

PPT

Section 2 Primary Standards and Standard Solutions

1. Primary standard

A primary standard (基准物质) is a highly purified compound that serves as a reference material in titrations and in other analytical methods. A primary standard can be used for the preparation and standardization of a standard solution.

1.1 Conditions for primary standard

A chemical compound must fulfill several requirements in order to serve as a satisfactory primary standard.

(1) The compound should be highly pure. It is pure enough to be weighed and used directly: 99.9% pure, or better.

(2) The composition of compound completely accords with the chemical formula, including the number of crystal water.

(3) The compound should stay stable. It should not decompose under ordinary storage. It should not absorb water or carbon dioxide from the atmosphere. And it should be stable when dried by heat or vacuum, because drying is required to remove traces of water adsorbed from the atmosphere.

(4) The compound should, if possible, have a large molar mass, because weighing out a greater weight of material tends to minimize the relative error caused by weighing.

1.2 Common primary standard

Very few compounds meet these criteria, and only a limited number of primary standards are available commercially. The primary standards commonly used are shown in Table 4-1.

Table 4-1 Drying methods and application objects of common primary standards

Primary standard	Drying method	Dried composition	Application object
$Na_2B_4O_7 \cdot 10H_2O$	Storage in a dryer containing an aqueous solution saturated with NaCl and sucrose	$Na_2B_4O_7 \cdot 10H_2O$	Acid
Na_2CO_3	270–300℃	Na_2CO_3	Acid
$KHC_8H_4O_4$ (potassium hydrogen phthalate)	105–110℃	$KHC_8H_4O_4$	Base or $HClO_4$
$H_2C_2O_4 \cdot 2H_2O$	Storage in atmosphere at room temperature	$H_2C_2O_4 \cdot 2H_2O$	Base or $KMnO_4$
$Na_2C_2O_4$	130℃	$Na_2C_2O_4$	Oxidant

continued

Primary standard	Drying method	Dried composition	Application object
$K_2Cr_2O_7$	140–150℃	$K_2Cr_2O_7$	Reductant
As_2O_3	Storage in a dryer at room temperature	As_2O_3	Oxidant
KIO_3	130℃	KIO_3	Reductant
ZnO	800℃	ZnO	EDTA
$CaCO_3$	110℃	$CaCO_3$	EDTA
Zn	Storage in a dryer at room temperature	Zn	EDTA
NaCl	500–600℃	NaCl	$AgNO_3$
$AgNO_3$	280–290℃	$AgNO_3$	Chloride

2. Standard solution

A standard solution is a reagent of known concentration. Standard solutions are used in titrations and in many other chemical analyses. The concentration of a standard solution should be sufficiently stable in order to ensure the accuracy of analysis results. Standard solutions play a central role in all titrations. Therefore, we must consider the desirable properties for such solutions, how they are prepared, and how their concentrations are expressed.

2.1 Preparation of standard solution

Two basic methods are used to prepare a standard solution according to the different properties of compounds.

2.1.1 Direct compounding method

The first way to prepare a standard solution is the direct method in which a primary standard with a certain amount is dissolved in a suitable solvent and diluted to an exactly known volume in a volumetric flask. The accurate concentration of a standard solution can be calculated by the mass of a primary standard and the volume of a volumetric flask. This direct method is simple, but only a primary standard can be used.

2.1.2 Indirect compounding method

Many reagents used as titrants, such as HCl, are not available as primary standards. Instead, indirect method is used to prepare the titrant. The indirect compounding method is to prepare a solution with approximately the desired concentration and then to titrate an analyte that is a primary standard or another standard solution. By this procedure, called standardization.

A titrant that is standardized is sometimes referred to as a secondary-standard solution. The concentration of a secondary-standard solution is subject to a larger uncertainty than the concentration of a primary-standard solution. If there is a choice, then, solutions are best prepared by the direct method. Many reagents, however, lack the properties required for a primary standard and, therefore, require preparation using indirect method and standardization.

2.2 Standardization of standard solution

The titration procedures of determining the concentration of a standard solution with primary standard or another standard solution is called standardization (标定).

2.2.1 Standardization with a primary standard

The concentration of a standard solution is determined by titrating it with a primary standard. The primary standard should be weighed an accurate mass, dissolved in a suitable solvent and diluted. The accurate concentration of a standard solution can be calculated by the mass of a primary standard and the volume of a standard solution. For example, a standard solution of purified sodium hydroxide is prepared, and this is standardized by titrating accurately weighed portions of a primary standard acid, such as potassium acid phthalate.

2.2.2 Compared with a standard solution

A known volume of the standard solution is determined by titrating it with an exactly known volume of another standard solution. The accurate concentration of a standard solution can be calculated by the volume of a standard solution and the volume and concentration of another standard solution. In this standardization, the concentration of a standard solution is determined by comparing with another standard solution.

2.3 The concentration expression way of standard solution

Amount-of-substance concentration or titer are the concentration expression ways of a standard solution usually used in the titrations.

2.3.1 Amount-of-substance concentration

The amount-of-substance concentration of a solution is the number of moles of solute present in one liter of solution.

$$c = \frac{n}{V} \tag{4-1}$$

Where n is the amount of the solute (mol or mmol), V is the volume of the solution (L or ml), and c is the molar concentration (mol/L or mmol/L).

A mole is the formula weight of a substance express in grams:

$$n = \frac{m}{M} \tag{4-2}$$

Where m is the mass of the solute (g), and M is the formula weight or molar mass of the solute (g/mol).

2.3.2 Titer

The titer (滴定度) is the mass of analyte that is chemically equivalence to 1ml of the titrant. The titer is usually expressed as $T_{T/A}$, where T and A are the chemical formulas of titrant and analyte, respectively, and the unit of titer is g/ml or mg/ml. In routine titrations, it is convenient to obtain the mass of the analyte from titer according to the following relation, where m_A is the mass of the analyte, and V_T is the volume of the titrant required in titration.

$$m_A = T_{T/A} \times V_T \tag{4-3}$$

For instance, the expression $T_{K_2Cr_2O_7/Fe} = 0.005000\text{g/ml}$ implies that 1ml of the standard solution $K_2Cr_2O_7$ is equivalent to 0.005000g of iron. If the volume of standard solution $K_2Cr_2O_7$ consumed in titration is 21.02ml, the mass of Fe can be calculated as followings.

$$m_{Fe} = T_{K_2Cr_2O_7/Fe} \times V_{K_2Cr_2O_7} = 0.005000 \times 21.02 = 0.1051\text{(g)}$$

Titer is frequently used in Chinese Pharmacopoeia, so that it is an important calculation way in pharmaceutical analysis.

PPT

Section 3 The Calculations in Titrimetric Analysis

1. The calculation basis of titrimetric analysis

In a direct titration, the relationship between titrant and analyte is conform to the stoichiometric ratio of reaction. This is the calculation basis of titrimetric analysis. If the volume and concentration of the titrant are known, the unknown quantity of the analyte can be calculated. The relation between the titrant T and analyte A is as follows:

$$tT + aA \rightleftharpoons bB + cC$$

Where T is the titrant, A is the analyte, and t and a are the numbers of moles of each.

At the stoichiometric point, t mol T and a mol A react completely. The mole ratio of titrant and analyte can be expressed as follows.

$$n_T : n_A = t : a$$

$$n_T = \frac{t}{a} n_A \quad \text{or} \quad n_A = \frac{a}{t} n_T \qquad (4\text{-}4)$$

2. Related calculation of titration analysis method

Equations 4-1, 4-2 and 4-4 are the basic relationship in titrimetric calculations. Here are some examples to illustrate stoichiometry calculations.

2.1 The calculation of standard solution concentration

2.1.1 Direct compounding method

【Example 4-1】

The 500.0ml of standard solution $K_2Cr_2O_7$ (0.01000mol/L) is required to prepare for oxidation-reduction titration. How many grams of primary standard $K_2Cr_2O_7$ should be weighed? ($M_{K_2Cr_2O_7} = 294.18$g/mol)

Solution:

$$m_{K_2Cr_2O_7} = n \cdot M = c \cdot V \cdot M = 0.01000 \times 500.0 \times 10^{-3} \times 294.18 = 1.471(\text{g})$$

2.1.2 Indirect compounding method (Including the calculation of solution dilution and concentrated)

The number of moles in a concentrated solution must equal to the number of moles in the dilute solutions. The amount of the solute is independent of solution dilute or concentrated, only but the molar concentration and the volume change.

$$n_{\text{concd}} = n_{\text{dil}} \qquad (4\text{-}5)$$

【Example 4-2】

The concentration of concentrated sulfuric acid is about 18mol/L. If you would prepare 1000ml sulfuric acid solution (0.10mol/L) from concentrated sulfuric acid, then how many milliliters concentrated sulfuric acid should be diluted?

Solution: $n_{\text{concd}} = n_{\text{dil}}$ $\qquad c_{\text{concd}} \cdot V_{\text{concd}} = c_{\text{dil}} \cdot V_{\text{dil}}$

$$V_{concd} = \frac{c_{dil} \cdot V_{dil}}{c_{concd}} = \frac{0.10 \times 1000}{18} \approx 5.6 \text{ (ml)}$$

In a standardization, the concentration of a standard solution is determined by a primary standard or another standard solution. The stoichiometric ratio of reaction is the calculation basis in a standardization.

【Example 4-3】

Titration of 0.2121g of primary standard $Na_2C_2O_4$ required 43.31ml of $KMnO_4$. What is the molar concentration of the $KMnO_4$ solution? ($M_{Na_2C_2O_4}$ = 134.00g/mol)

Solution: The titrimetric reaction is

$$2MnO_4^- + 5C_2O_4^{2-} + 16H^+ \rightleftharpoons 2Mn^{2+} + 10CO_2\uparrow + 8H_2O$$

$$n_{KMnO_4} : n_{Na_2C_2O_4} = 2 : 5$$

$$c_{KMnO_4} = \frac{n_{KMnO_4}}{V_{KMnO_4}} = \frac{\frac{2}{5} \cdot n_{Na_2C_2O_4}}{V_{KMnO_4}} = \frac{\frac{2}{5} \times \frac{m_{Na_2C_2O_4}}{M_{Na_2C_2O_4}}}{V_{KMnO_4}} = \frac{\frac{2}{5} \times \frac{0.2121}{134.00}}{43.31 \times 10^{-3}} = 0.01462(\text{mol/L})$$

【Example 4-4】

A 50.00ml portion of an HCl solution required 29.71ml of 0.01963mol/L $Ba(OH)_2$ to reach an end point with bromocresol green indicator. Calculate the molar concentration of the HCl.

Solution: The titrimetric reaction is

$$Ba(OH)_2 + 2HCl = BaCl_2 + 2H_2O$$

$$n_{Ba(OH)_2} : n_{HCl} = 1 : 2$$

$$c_{HCl} = \frac{n_{HCl}}{V_{HCl}} = \frac{2n_{Ba(OH)_2}}{V_{HCl}} = \frac{2 \cdot c_{Ba(OH)_2} \cdot V_{Ba(OH)_2}}{V_{HCl}} = \frac{2 \times 0.01963 \times 29.71 \times 10^{-3}}{50.00 \times 10^{-3}} = 0.02333(\text{mol/L})$$

2.2 Conversion between amount-of-substance concentration and titer

When the volume of the titrant V_T is regard as 1ml, $T_{T/A}$ is equal to the mass of analyte m_A according to the define of titer. So that, the relation between the molar concentration and titer is as follows.

$$\frac{c_T \times 10^{-3}}{T_{T/A}/M_A} = \frac{n_T}{n_A} = \frac{t}{a}$$

$$c_T = \frac{t}{a} \cdot \frac{T_{T/A} \times 10^3}{M_A} \quad \text{or} \quad T_{T/A} = \frac{a}{t} \cdot \frac{c_T M_A}{10^3} \tag{4-6}$$

【Example 4-5】

Calculate the titer of a 0.1043mol/L standard HCl solution in $T_{HCl/CaCO_3}$. (M_{CaCO_3} = 100.1g/mol)

Solution: The titrimetric reaction is

$$CaCO_3 + 2HCl = CaCl_2 + H_2O + CO_2\uparrow$$

$$n_{CaCO_3} : n_{HCl} = 1 : 2$$

$$T_{HCl/CaCO_3} = \frac{1}{2} \times \frac{0.1043 \times 100.1}{10^3} = 0.005220 \text{ (g/ml)}$$

2.3 The calculation of the content of the substance to be measured

In titrimetric analysis, mass fraction is often used to express the content of the analyte. Surposing that the mass of the sample taken is m_s and the mass of the analyte in the sample is found to be m_A, then the mass fraction ω_A of the analyte in the sample may be expressed as:

$$\omega_A = \frac{m_A}{m_s} \times 100\% = \frac{\frac{a}{t} \cdot c_T \cdot V_T \cdot M_A}{m_s} \times 100\% \quad (4\text{-}7)$$

【Example 4-6】

A 0.1250g sample of drug Na_2CO_3 was titrated with 23.00ml of 0.1005mol/L HCl standard solution. Calculate the mass fraction of Na_2CO_3 in the drug. ($M_{Na_2CO_3}$ = 105.99g/mol)

Solution: The titrimetric reaction is

$$Na_2CO_3 + 2HCl = 2NaCl + H_2O + CO_2\uparrow$$

$$n_{Na_2CO_3} : n_{HCl} = 1 : 2$$

$$\omega_{Na_2CO_3} = \frac{m_{Na_2CO_3}}{m_s} \times 100\% = \frac{\frac{1}{2} \times 0.1005 \times 23.00 \times 10^{-3} \times 105.99}{0.1250} \times 100\% = 98.00\%$$

【Example 4-7】

A 0.6501g sample of drug in which contains S. $Na_2S_2O_3$ can be quantitative displaced from the reaction between S and Na_2SO_3, and then can be direct titrated with 21.60ml titrant I_2. Calculate the mass fraction of S in the sample. ($T_{I_2/S}$ = 3.142mg/ml).

Solution:

$$\omega_S = \frac{m_A}{m_s} \times 100\% = \frac{T_{I_2/S} \cdot V_{I_2}}{m_s} \times 100\% = \frac{3.142 \times 10^{-3} \times 21.60}{0.6501} \times 100\% = 10.44\%$$

【Example 4-8】

A 0.1983g $CaCO_3$ sample was completely dissolved in 25.00ml of 0.2010mol/L HCl standard solution. The HCl remaining after reaction with $CaCO_3$ can then be back titrated with 9.50ml of 0.2000mol/L NaOH standard solution. Calculate mass fraction of $CaCO_3$ in the sample. (M_{CaCO_3} = 100.1g/mol)

Solution: The titrimetric reaction is

$$CaCO_3 + 2HCl\ (excess) = CaCl_2 + H_2O + CO_2\uparrow$$

$$HCl\ (remaining) + NaOH = NaCl + H_2O$$

$$n_{CaCO_3} : n_{HCl} = 1 : 2 \qquad n_{NaOH} : n_{HCl} = 1 : 1$$

$$\omega_{Ca_2CO_3} = \frac{m_{CaCO_3}}{m_s} \times 100\% = \frac{\frac{1}{2} \cdot n_{HCl} \cdot M_{CaCO_3}}{m_s} \times 100\%$$

$$= \frac{\frac{1}{2} \times (0.2010 \times 25.00 \times 10^{-3} - 0.2000 \times 9.50 \times 10^{-3}) \times 100.1}{0.1983} \times 100\% = 78.87\%$$

重点小结

滴定分析法（titrimetric analysis），也称为“容量分析法”（volumetric analysis），是一种将标准溶液（standard solution）滴加到待测物溶液中，直到化学计量点，然后根据标准溶液所消耗浓度和体积，计算待测组分含量的分析方法。化学计量点（stoichiometric point，sp），是指当加入的标准

溶液的量与待测组分的量之间按化学反应的计量关系恰好反应完全的点。滴定终点（end-point of titration，ep）是指在滴定过程中溶液的颜色或电位、电导、光度等仪器信号发生突变的点。滴定误差（titration error，TE）又称终点误差（end point error）是指滴定终点与化学计量点不一致所引起的分析误差。

滴定分析对滴定反应的三个基本要求：①反应必须定量完成。包括反应有确定的化学计量关系和反应完全。②反应速度要快。③有适当的终点指示方法。同时满足以上三个要求可用于直接滴定法，为了扩展滴定分析的应用范围，还可以采用返滴定法、置换滴定法和间接滴定法。

基准物质应具备的条件为：纯度高、组成与化学式完全符合、性质稳定、最好具有较大的摩尔质量。标准溶液的配制方法包括直接配制法和间接配制法。必须是基准物质才能用直接配制法，其他应采用间接配制法再标定其准确浓度。标准溶液的标定方法有用基准物质标定和用另一种标准溶液比较法标定。标准溶液浓度的表示方法有物质的量浓度和滴定度，滴定度（titer）是指每毫升滴定剂相当于被测物质的质量，用 $T_{T/A}$ 表示。

在滴定分析中，滴定剂和被测组分的化学反应计量关系是滴定分析的依据，以此为基础计算滴定分析中被测组分的含量。

题库

目标检测

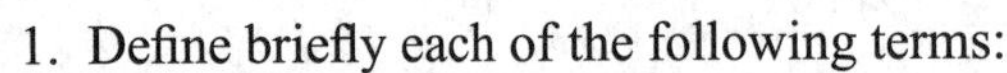

1. Define briefly each of the following terms:

(a) standard solution (b) titration (c) standardization (d) titer

2. Distinguish between the stoichiometric point and the end point of a titration.

3. Which of the following substances can be prepared as a standard solution by the direct compounding method? Why is that?

$$NaOH, HCl, Na_2B_4O_7 \cdot 10H_2O, Na_2C_2O_4, KMnO_4, K_2Cr_2O_7, NaCl, Na_2S_2O_3$$

4. What is the molar concentration of NaCl when 3.2580g are dissolved in water and diluted to 500.0ml? (M_{NaCl} = 58.440g/mol)

5. The density of 37% (*m*/*m*) concentrated hydrochloric acid is 1.19kg/L. Calculate the molar concentration of concentrated hydrochloric acid. If you would prepare 1L HCl solution (0.10mol/L) from concentrated hydrochloric acid, then how many milliliters concentrated hydrochloric acid should be diluted? (M_{HCl} = 36.458g/mol)

6. Titration of 0.5342g of primary standard $Na_2B_4O_7 \cdot 10H_2O$ required 27.98ml of HCl solution. What is the molar concentration of the HCl solution? ($M_{Na_2B_4O_7 \cdot 10H_2O}$ = 381.36g/mol)

7. The titer of a standard HCl solution is $T_{HCl/CaO}$ = 0.005879g/ml. Calculate the molar concentration of HCl solution. (M_{CaO} = 56.077g/mol)

8. A 0.5005g sample of an iron ore is dissolved in acid. The iron is then reduced to Fe^{2+} and titrated with 25.12ml of $K_2Cr_2O_7$ standard solution. If the titer of $K_2Cr_2O_7$ standard solution is $T_{K_2Cr_2O_7/Fe}$ = 0.005314g/ml. Calculate the mass fraction of Fe in the sample.

9. Titration of the I_2 produced from 0.1142g of primary standard KIO_3 required 27.95ml of sodium thiosulfate. Calculate the concentration of the $Na_2S_2O_3$. (M_{KIO_3} = 214.00g/mol)

$$IO_3^- + 5I^- + 6H^+ \rightleftharpoons 3I_2 + 3H_2O$$

$$I_2 + 2S_2O_3^{2-} \rightleftharpoons 2I^- + S_4O_6^{2-}$$

10. A 0.4126g sample of primary standard Na_2CO_3 was treated with 40.00ml of dilute perchloric acid. The solution was boiled to remove CO_2, following which the excess $HClO_4$ was back-titrated with 9.20ml of dilute NaOH. In a separate experiment, it was established that 26.93ml of the $HClO_4$ neutralized the NaOH in a 25.00ml portion. Calculate the molarity of the $HClO_4$. ($M_{Na_2CO_3}$ = 105.99g/mol)

Chapter 5　Acid-Base Titrations

学习目标

知识要求

1. **掌握**　酸碱滴定法的基本原理，包括溶液中的酸碱平衡理论，溶液中氢离子浓度及 pH 的计算，化合物能否被准确滴定的判断，指示剂的变色原理、变色范围，酸碱滴定曲线以及指示剂的选择，终点误差的计算。

2. **熟悉**　有关滴定反应过程的基本概念，包括滴定终点、滴定曲线等；常用的酸性标准溶液和碱性标准溶液。

3. **了解**　酸碱滴定的应用；非水滴定的原理和应用。

能力要求

通过本章的学习，掌握滴定分析法的基本理论和相关的实验技能。

PPT

Section 1　Introduction of Acid-Base Titrations

Acid-base titration is a titration method based on the reaction of an acid and a base, which is one of the most important methods among titration methods and characterized by easy and fast. It is widely used to detect various kinds of acids, bases and substances that can react with an acid or a base along with proton transfer directly or indirectly. So, it is common to apply this method to analyze the traditional Chinese medicine, chemical synthetic drugs, and biological samples.

Usually, there is no significant apparent change at the stoichiometric points for acid-base reactions, therefore indicators and instruments are needed for determining the end points of titrations. It is an easy and convenient method that determines the end points of titrations with the help of the color change of indicators, resulted the widely application in practice. The key points of acid-base titration are to estimate whether the analytes can be titrated accurately and to choose the appropriate indicators which can determine the end points, and these depend on the change of pH during titrating. Hence, for discussing the acid-base titration, it is necessary to know the pH variations of solutions during titrating, the principle of discoloration and discoloration range of acid-base indicators and the principle of acid-base indicators selection.

To master acid-base titration, first, the essence of the acid-base reaction and the theories of acid-base equilibrium must be understood. This chapter applies acid-base proton theory to solve the questions about acid-base equilibrium and unify acid-base equilibrium in aqueous solution and nonaqueous solution. Base on the calculation of $[H^+]$, discuss the acid-base equilibrium in aqueous solution, the calculation of components concentration in the equilibrium system, and the theories and practical applications of acid-base titration.

PPT

Section 2 Acid-Base Equilibria in Aqueous Solutions

Given that the titration in this chapter is based upon aqueous solution and acid-base reaction taking place in it, the properties of water and some basic theories of acid-base reaction are necessarily introduced in following topics before we consider the practical aspects of the titrations.

1. Acid-base proton theory

1.1 The definition of an acid and a base

According to the Bronsted-Lowry acid-base theory, an acid is a molecule which is capable of acting as a proton donor, while a base, in turn, is a molecule which is capable of acting as a proton acceptor. After an acid undergoes dissociation and ionization in aqueous solution, it donates a proton and generates its conjugate base:

$$\underset{\text{acid}}{HA} \rightleftharpoons \underset{\text{base}}{A^-} + H^+$$

Such two species differing from each other only by a proton are called a conjugate acid-base pair. In this context, acids and bases are not supposed to be strictly neutral, they may also be anions and cations, like examples below:

$$HAc \rightleftharpoons Ac^- + H^+$$

$$NH_4^+ \rightleftharpoons NH_3 + H^+$$

$$HS^- \rightleftharpoons S^{2-} + H^+$$

$$^+H_3N—R—NH_3^+ \rightleftharpoons H_3N^+—R—NH_2 + H^+$$

Charges do not matter in the determination of acids or bases. Some species can either donate or accept protons, like HS^-, which allow them to act as either acid or base according to the specific reaction.

1.2 The concept of solvent proton and the essence of acid-base reaction

Fundamentally, water can either donate or accept protons, namely it is both a Bronsted-Lowry acid and Bronsted-Lowry base. And moreover, as the protons or hydrogen ions H^+ cannot exist freely in solution, they greatly tend to combine with one or more molecules of water to form the hydronium ions, illustrated by H_3O^+. Subsequently, for example, when an acid dissociates in aqueous solution, the proton-transfer actually proceeds between the parent acid and water, so another expression for the hydrochloric acid dissociation in water under the scheme of equation 5-1 is:

$$HCl\ (aq) + H_2O\ (l) \rightleftharpoons H_3O^+\ (aq) + Cl^-\ (aq)$$

Then the essence of acid-base reaction (also known as neutralization), can be defined as the transfer of protons from the acid molecule to the base molecule. Therefore, after we add up the general equation

$acid_n \rightleftharpoons base_n$ + proton or $base_n$ + proton $\rightleftharpoons acid_n$ of two species, the reaction will be written as:

$$acid_1 + base_2 \rightleftharpoons acid_2 + base_1$$

1.3 The proton self-recursive constant of the solvent

Pure water is slightly self-ionized in nature, which is called autoprotolysis (质子自递).

$$H_2O + H_2O \rightleftharpoons H_3O^+ + OH^-$$

According to the law of chemical equilibria, the equilibrium constant K for the equilibria above is:

$$K = \frac{[H_3O^+][OH^-]}{[H_2O]^2}$$

The concentration of undissociated water remains virtually constant in dilute solutions (~55.5mol/L). For simplification, we use $[H^+]$ in place of $[H_3O^+]$. So the ionic product of water, represented by K_w, is:

$$K_w = [H^+]\,[OH^-] \tag{5-1}$$

K_w is a constant which only changes with temperature, and is independent of the concentration of solutes. In view of the common temperature condition in the practice of experiments and the convenience of calculation, the value of $K_w = 1\times10^{-14}$ at 25°C is always taken by chemists in titration calculation, as well as throughout this chapter.

At 25°C,

$$K_w = [H^+]\,[OH^-] = 1\times10^{-14}$$

$$pK_w = pH + pOH = 14.00$$

1.4 The relationship between the dissociation constant of conjugate acid-base pairs

The conjugate base of an acid is able to accept protons, whereas the conjugate acid of a base is able to give protons. This phenomenon is called the hydrolysis of the salt ion. Since the dissociation constant K of a weak acid or base is rather small because it just partially dissociates in the water, in other words, the dissociation equilibrium is not complete to the right and the undissociated form can still exist to some extent, the contribution of salt ion hydrolysis cannot be as negligible as strong acid or base. Here we take the dissociation of a weak acid HA in aqueous solution as an example:

$$HA \rightleftharpoons A^- + H^+$$

The acid equilibrium constant K_a is:

$$K_a = \frac{[H^+][A^-]}{[HA]}$$

And its conjugate base A^- will be hydrolyzed and generate hydroxide ion and HA:

$$A^- + H_2O \rightleftharpoons HA + OH^-$$

This equilibrium can be regarded the same as the weak base dissociation. And the base equilibrium constant K_b is:

$$K_b = \frac{[HA][OH^-]}{[A^-]}$$

After multiplying the equilibrium constant of the weak acid and its conjugate base, we obtain:

$$K_a \cdot K_b = \frac{[H^+][A^-]}{[HA]} \cdot \frac{[HA][OH^-]}{[A^-]} = [H^+][OH^-] = K_w$$

$$K_a \cdot K_b = K_w \tag{5-2}$$

And,

$$pK_a + pK_b = pK_w \tag{5-3}$$

That is to say the conjugate base (acid) has a strength that vary inversely as the strength of the parent weak acid (base): the base conjugated to a weaker acid is relatively stronger, and the base conjugated to a stronger acid is relatively weaker. Using this equation, we can calculate K_b for the conjugate base if we know K_a. For example, K_a for acetic acid is 1.8×10^{5}, so K_b for acetate ion is:

$$K_b = \frac{K_w}{K_a} = \frac{1.0 \times 10^{-14}}{1.8 \times 10^{5}} = 5.6 \times 10^{-10}$$

2. Distribution of components in an acid-base solution

2.1 Acid concentration, acidity and equilibrium concentration

Acid concentration is defined as the number of moles of acid (including both undissociated and dissociated form) per unit volume, which is commonly represented by *c*. Acidity, often indicated with pH, is the hydrogen ion activity in the solution relating to the types and concentrations of acids. These two concepts are distinct with each other. Similarly, for bases, their basicity is indicated with pOH.

For a weak acid, the imcomplete dissociation results in more than one species (there will be even more for a polyprotic system) presenting in the solution at the equilibrium. The equilibrium concentration of one species is expressed by a bracket ([ ]). The total equilibrium concentration of a weak acid or base is obtained by adding the equilibrium concentration of every single species.

In the case of 0.1000mol/L HAc aqueous solution, only 0.001340mol/L of HAc dissociates into Ac^-. The equilibrium concentrations of HAc and Ac^- are given as:

$$HAc \rightleftharpoons Ac^- + H^+$$

$$c_{HAc} = [HAc] + [Ac^-]$$

$$[Ac^-] = 0.001340\text{mol/L}$$

$$[HAc] = c_{HAc} - [Ac^-] = 0.1000 - 0.001340 = 0.09866\text{mol/L}$$

2.2 The distribution coefficient of the acid-base

For the dissociation of acetic acid in water, there are two species coexisting, namely HAc and Ac^-:

$$HAc \rightleftharpoons Ac^- + H^+$$

$$K_a = \frac{[H^+][Ac^-]}{[HAc]}$$

The total concentration of acid is the sum of the concentration of both HAc and Ac^-:

$$c_{HAc} = [HAc] + [Ac^-]$$

To indicate the fraction of undissociated weak acid and its conjugate base, we define the distribution coefficient δ_i of species as:

$$\delta_{HAc} = \frac{[HAc]}{c_{HAc}} = \frac{[HAc]}{[HAc]+[Ac^-]} = \frac{1}{1+[Ac^-]/[HAc]} = \frac{1}{1+K_a/[H^+]} = \frac{[H^+]}{[H^+]+K_a}$$

$$\delta_{HAc} = \frac{[H^+]}{[H^+]+K_a} \tag{5-4}$$

Substitute HAc with Ac^- and calculate in the same way, we get:

$$\delta_{Ac^-} = \frac{[Ac^-]}{c_{HAc}} = \frac{[Ac^-]}{[HAc]+[Ac^-]} = \frac{1}{1+[HAc]/[Ac^-]} = \frac{1}{1+[H^+]/K_a} = \frac{K_a}{[H^+]+K_a} \tag{5-5}$$

Apparently,

$$\delta_{Ac-} + \delta_{HAc} = 1$$

Because K_a is a constant under a given temperature, so δ_i is a function of $[H^+]$. If we draw plots of δ_i and pH based on the relationship above, a distribution curve will be obtained, and a collection of those distribution curves of different species will form the distribution diagram of that acid or base solution.

Using distribution diagrams, the change in composition of species in the solution is clearly showed with the change of pH, like the distribution diagram of acetic acid on the left. When pH is below pK_a, $[Ac^-] < [HAc]$, HAc is the dominant form in the solution; at the cross-over point, where pH = pK_a and $[HAc] = [Ac^-]$, the solution is a 1 : 1 mixture of the two forms; when pH is above pK_a, $[Ac^-] > [HAc]$, Ac^- is the dominant form in the solution.

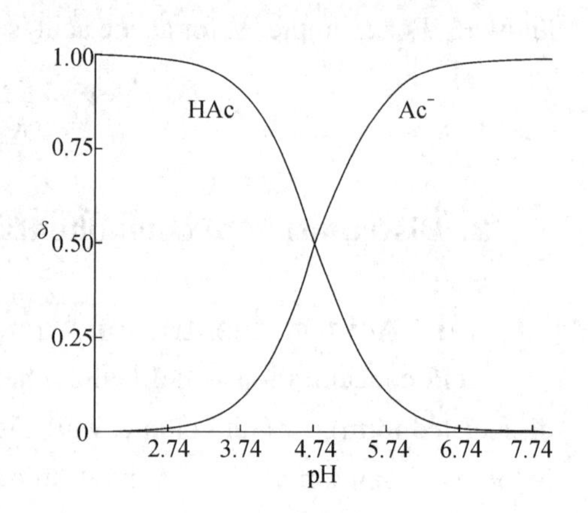

Figure 5-1 δ-pH curve of HAc

The estimation of the major species according to such three pH regions is also true for other monoprotic weak acids and bases. And when a pH is given, we could always get the fraction of species from $[H^+]$.

【Example 5-1】

For weak acid HAc, calculate the δ_{HAc}, δ_{Ac^-}, [HAc], and $[Ac^-]$ of a 0.1000mol/L solution at pH = 5.00.

Solution:

$$K_{a(HAc)} = 1.8 \times 10^{-5}, \text{ and at pH} = 5.00, [H^+] = 1.0 \times 10^{-5}$$

$$\delta_{HAc} = \frac{[H^+]}{[H^+] + K_a} = \frac{1.0 \times 10^{-5}}{1.0\times10^{-5}+1.8\times10^{-5}} = 0.36$$

$$\delta_{Ac^-} = 1-\delta_{HAc} = 1-0.36 = 0.64$$

$$[HAc] = \delta_{HAc} \cdot c_{HAc} = 0.36 \times 0.1000 = 3.6 \times 10^{-2} \text{(mol/L)}$$

$$[Ac^-] = \delta_{Ac^-} \cdot c_{Ac^-} = 0.64 \times 0.1000 = 6.4 \times 10^{-2} \text{(mol/L)}$$

For a monoprotic weak base, the reasoning of the concentration calculation and distribution diagram construction is exactly the same.

3. Calculation of $[H^+]$ in an acid-base solution

3.1 Proton balance equation

According to the proton theory, the acid-base reaction is the proton transfer reaction, and in most cases the solvent molecules are also involved in this process. Therefore, when dealing with the balance problem of acid-base reaction in solution, the solvent should be considered. The chemical equilibrium in acid-base solution should consider from mass equilibrium, charge equilibrium and proton equilibrium. This is the fundamental basis for studying solution equilibrium.

Mass balance (material balance) means that the total concentration of a given component in a chemical equilibrium system should be equal to the sum of the equilibrium concentration of its related components. This equilibrium relation is called mass equilibrium and its mathematical expression is called mass (material) balance equation (质量平衡式). For example, the mass balance equation of c mol/L HAc aqueous solution is:

$$c_{HAc} = [HAc] + [Ac^-]$$

The mass balance equation of c mol/L Na_2CO_3 aqueous solution is:

$$c = [CO_3^{2-}] + [HCO_3^-] + [H_2CO_3]$$
$$c = [Na^+]/2$$

Therefore, the mass balance equation indicates the relationship between equilibrium concentration and analytical concentration.

Charge balance means that in a chemical equilibrium system, the total charge of the cations in the solution must equal the total charge of the anions, including H^+ and OH^- from the dissociation of the solvent water, and the solution is always electrically neutral. This equilibrium relation is called charge balance and its mathematical expression is called charge balance equation. For example, the charge balance equation of *c* mol/L HCN solution is:

$$[H^+] = [CN^-] + [OH^-]$$

The charge balance equation of *c* mol/L NaH_2PO_4 solution is:

$$[Na^+] + [H^+] = [H_2PO_4^-] + 2[HPO_4^{2-}] + 3[PO_4^{3-}] + [OH^-]$$

In the formula, the coefficient 2 before $[HPO_4^{2-}]$ means that each HPO_4^{2-} has two negative charges, and $3[PO_4^{3-}]$ is similar. Obviously, the neutral molecules do not participate in the charge balance equation.

Proton equilibrium means that the total number of protons lost by the acid must equal the total number of protons accepted by the base when the acid-base reaction reaches an equilibrium. So the protons total number that proton-acceptors accept should be equal to the protons total number that proton-donors lose. The equal-transverse relationship of proton transfer between acid and base is called proton equilibrium or proton condition, and its mathematical expression is called proton balance equation (质子条件式). According to the proton balance equation, we can obtain the relation between the concentration of H^+ in solution and the concentration of related components, which is the fundamental equations to deal with the calculation of acid-base equilibria.

Generally, the substance (initial solute and solvent) that is sufficient in the solution and participates in proton transferring reactions is selected as the zero level (proton reference benchmark) to judge the products after donating or accepting protons, and the number of transfer protons. And, writing the proton balance equation bases on the principle that that the number of protons donated and accepted must be equal.

For example, HAc aqueous solution chooses HAc and H_2O as the zero level, and its proton transfer reaction is:

$$HAc \rightleftharpoons H^+ + Ac^-$$
$$H_2O \rightleftharpoons H^+ + OH^-$$

H^+ (or H_3O^+) is the product from proton acceptor, and OH^- and Ac^- are the product from proton donator.

The proton balance equation is: $[H^+] = [Ac^-] + [OH^-]$

【Example 5-2】

Write the proton balance equation of Na_2HPO_4 aqueous solution.

Solution:

HPO_4^{2-} and H_2O are the zero levels, and its proton transfer reaction is:

$$HPO_4^{2-} \rightleftharpoons H^+ + PO_4^{3-}$$
$$HPO_4^{2-} + H_2O \rightleftharpoons H_2PO_4^- + OH^-$$
$$H_2PO_4^- + H_2O \rightleftharpoons H_3PO_4 + OH^-$$
$$H_2O \rightleftharpoons H^+ + OH^-$$

From the above equilibrium, PO_4^{3-} and OH^- are the products that donate proton and H^+, $H_2PO_4^-$, H_3PO_4 are the products that accept proton. But it should be noted that H_3PO_4 is the product of HPO_4^{2-} by accepting two protons, so in order to equal the number of protons donated and accepted, its concentration should be multiplied by the coefficient 2. The proton balance equation of Na_2HPO_4 aqueous solution is:

$$[H^+] + [H_2PO_4^-] + 2[H_3PO_4] = [PO_4^{3-}] + [OH^-]$$

The above proton balance equation considers both the dissociation of solute and the dissociation of solvent, so the proton balance equation shows the rigorous quantitative relationship in the acid-base equilibria system.

The proton balance equation of the acid-base equilibria system can also be derived from the charge equilibrium equation and the mass equilibrium equation. For example, c mol/L $NaHCO_3$ aqueous solution:

The mass balance equation is:

$$c = [H_2CO_3] + [HCO_3^-] + [CO_3^{2-}] \quad \text{(a)}$$

The proton balance equation is: $[Na^+] + [H^+] = [HCO_3^-] + 2[CO_3^{2-}] + [OH^-]$

Because $[Na^+]=c$,

$$[H^+] + c = [HCO_3^-] + 2[CO_3^{2-}] + [OH^-] \quad \text{(b)}$$

Unite (a) and (b):

$$[H^+] = [CO_3^{2-}] + [OH^-] - [H_2CO_3] \quad \text{(c)}$$

(c) is the proton balance equation of $NaHCO_3$ aqueous solution.

3.2 The calculation of $[H^+]$ in monoprotic acid (base) solution

3.2.1 The calculation of $[H^+]$ in monoprotic strong acid (strong base) solution

Strong acids and bases are completely dissociated in solution. Therefore, in general, the calculation of the acidity is relatively simple. In c mol/L strong acid HA aqueous solution, the following dissociation equilibrium exists:

$$HA \rightleftharpoons H^+ + A^-$$

$$H_2O \rightleftharpoons H^+ + OH^- \quad K_S^{H_2O} = [H^+][OH^-]$$

The proton balance equation is: $[H^+]=[A^-]+[OH^-]$

Since the strong acid is completely dissociated in the solution, then $[A^-]=c$, substitute into the proton condition: $[H^+] = c + [OH^-]$

When $c \geqslant 10^{-6}$mol/L, the dissociation of water is negligible, so

$$[H^+] = [A^-] = c \quad \text{(5-6)}$$

For example, a 0.1mol/L HCl aqueous solution, $[H^+] = 0.1$mol/L.

For strong bases solution, similar equations can be derived for calculation of $[OH^-]$, $[OH^-]=c$. For example, a 0.1mol/L NaOH aqueous solution, $[OH^-]=0.1$mol/L.

3.2.2 The calculation of $[H^+]$ in monoprotic weak acid (weak base) solution

In a c mol/L HA aqueous solution, the following dissociation equilibrium exists:

$$HA \rightleftharpoons H^+ + A^- \quad K_a = \frac{[H^+][A^-]}{[HA]}$$

$$H_2O \rightleftharpoons H^+ + OH^- \quad K_S^{H_2O} = [H^+][OH^-]$$

The proton balance equation is: $[H^+]=[A^-]+[OH^-]$

Substituting the expressions of the dissociation constants K_a、$K_S^{H_2O}$ into it:

$$[H^+]=\frac{K_a[HA]}{[H^+]}+\frac{K_S^{H_2O}}{[H^+]}$$

$$[H^+]=\sqrt{K_a[HA]+K_S^{H_2O}} \tag{5-7a}$$

above, $[HA]=\delta_{HA}\cdot c_{HA}=\frac{[H^+]}{[H^+]+K_a}\cdot c_{HA}$

Formula (5-7a) is the exact formula for calculating $[H^+]$ of a monoprotic weak acid solution. In practice, according to the allowable error in calculating acidity, the approximate calculation is allowed. When $K_a\cdot c\geqslant 20K_S^{H_2O}$, the dissociation of water can be neglected; and when $c/K_a\geqslant 500$, the acid is too weak, so the effect of its dissociated $[H^+]$ on the total concentration can be neglected, $[HA]=c-[H^+]\approx c$, and (5-7a) can be simplified to:

$$[H^+]=\sqrt{K_a\cdot c} \tag{5-7b}$$

Formula (5-7b) is the simplest formula to calculate the concentration of H^+ in a monoprotic weak acid solution.

The calculation of monoprotic weak base is in the same way. When $K_b\cdot c\geqslant 20K_S^{H_2O}$ and $c/K_b\geqslant 500$,

$$[OH^-]=\sqrt{K_b\cdot c} \tag{5-8}$$

【Example 5-3】

Calculate the pH of a weak acid HAc solution with concentration of 0.1000mol/L.

Solution:

$$K_a=1.8\times10^{-5} \quad c=0.10\text{mol/L}$$

Because of $K_a\cdot c>20K_S^{H_2O}$, $c/K_a>500$, it is reasonable to use the simplest formula.

$$[H^+]=\sqrt{K_a\cdot c}=\sqrt{1.8\times10^{-5}\times0.10}=1.3\times10^{-3}(\text{mol/L})$$

$$pH=-\lg[H^+]=2.89$$

3.3 The calculation of $[H^+]$ in polybasic acid (base) solution

Polybasic acids dissociate in steps in solution, which is a complex acid-base equilibria system. For example, the proton balance equation of c mol/L diprotic acid H_2A aqueous solution is as follows: $[H^+]=[HA^-]+2[A^{2-}]+[OH^-]$.

Substituting the expression of K_{a1}、K_{a2}、$K_S^{H_2O}$ into the above equation, we obtain:

$$[H^+]=\frac{[H_2A]K_{a1}}{[H^+]}+2\frac{[H_2A]K_{a1}K_{a2}}{[H^+]^2}+\frac{K_S^{H_2O}}{[H^+]}$$

Collated:

$$[H^+]=\sqrt{[H_2A]\,K_{a1}\left(1+\frac{2K_{a2}}{[H^+]}\right)+K_S^{H_2O}} \tag{5-9a}$$

This Formula (5-9a) is an exact formula for calculating $[H^+]$ of a diprotic weak acid aqueous solution.

Andintheaboveformula, $[H_2A]=\delta_{H_2A}\cdot c=\frac{[H^+]^2}{[H^+]^2+[H^+]K_{a1}+K_{a1}K_{a2}}\cdot c$

Generally, the dissociation constant of diprotic acid is $K_{a1}\gg K_{a2}\gg K_S^{H_2O}$. So when $K_{a1}\cdot c>20K_S^{H_2O}$, the dissociation of water can be neglected, and when $\frac{2K_{a2}}{[H^+]}\approx\frac{2K_{a2}}{\sqrt{K_{a1}\cdot c}}\leqslant 0.05$, the secondary dissociation can also be ignored. Therefore, this diprotic acid can be simplified as a monoprotic weak acid, and only the first step of dissociation should be considered, then $[H_2A]=c-[H^+]$; and when $c/K_{a1}\geqslant 500$, the dissociation of the diprotic acid is rather small, at this time, the equilibrium concentration of diprotic acid can be considered as its original concentration c, so (5-9a) can be simplified to:

$$[H^+] = \sqrt{K_{a1} \cdot c} \tag{5-9b}$$

The formula $[H^+] = \sqrt{K_{a1} \cdot c}$ is the simplest formula for calculating the $[H^+]$ of the diprotic acid aqueous solution. Other polybasic acids can also be treated in this way.

The calculation of $[OH^-]$ in polybasic weak base aqueous solution is the same as that of polybasic weak acid. The polybasic weak base are also dissociate step by step in solution, and its general law is $K_{b1} \gg K_{b2} \gg \ldots \gg K_S^{H_2O}$. So when $K_{b1} \cdot c \geqslant 20K_S^{H_2O}$ and $\frac{2K_{b2}}{\sqrt{K_{b1} \cdot c}} \leqslant 0.05$, polybasic base can be simplified as a monoprotic weak base, only the first step of its dissociation should be considered. And when $c/K_{b1} \geqslant 500$, the simplest formula is:

$$[OH^-] = \sqrt{K_{b1} \cdot c} \tag{5-10}$$

【Example 5-4】

Calculate the pH of saturated aqueous solution of H_2CO_3 (c=0.040mol/L) at room temperature.

Solution:

$K_{a1}=4.2\times10^{-7}$, $K_{a2}=5.6\times10^{-11}$, c=0.040mol/L

$$\because K_{a1} \cdot \frac{c}{K_S^{H_2O}} > 20, \frac{2K_{a2}}{[H^+]} \approx \frac{2K_{a2}}{\sqrt{K_{a1} \cdot c}} \leqslant 0.05,$$

$\therefore$ So we can use the simplest formula to calculate it:

$$[H^+] = \sqrt{K_{a1} \cdot c} = \sqrt{4.2\times10^{-7}\times0.0040} = 1.3\times10^{-4} \text{ (mol/L)}$$

$$pH=3.89$$

【Example 5-5】

Calculate the pH of aqueous solution of Na_2CO_3. (c=0.10mol/L)

Solution:

$$K_{b1} = \frac{K_S^{H_2O}}{K_{a2}} = 1.8\times10^{-4}, K_{b2} = \frac{K_S^{H_2O}}{K_{a1}} = 2.4\times10^{-8}$$

$$\because K_{b1} \cdot c/K_S^{H_2O} > 20, \frac{2K_{b2}}{\sqrt{K_{b1} \cdot c}} \leqslant 0.05, c/K_{b1} \geqslant 500$$

$\therefore$ So we can use the simplest formula to calculate it.

$$[OH^-] = \sqrt{K_{b1} \cdot c} = \sqrt{1.8\times10^{-4}\times0.10} = 4.2\times10^{-3} \text{ (mol/L)}$$

$$pOH=2.38 \qquad pH=11.62$$

3.4 The calculation of $[H^+]$ in amphoteric solution

Substances that act as both an acid and a base in a solution are called amphoteric substances (两性物质), such as HCO_3^-, $H_2PO_4^-$, HPO_4^{2-}, and so on. The calculation of $[H^+]$ in amphoteric solution is also according to acid-base equilibria. Write the proton balance equation, and the approximate value is calculated based on proton balance equation combining with the main balance in different situations.

There is taking NaHA hydroxide solution with the concentration of c as an example to explain the calculation of $[H^+]$ in amphoteric solution.

Proton balance equation:

$$[H^+] + [H_2A] = [OH^-] + [A^{2-}]$$

Substituting the expressions of each dissociation constant into it:

$$[H^+] + \frac{[H^+][HA^-]}{K_{a1}} = \frac{K_S^{H_2O}}{[H^+]} + \frac{K_{a2}[HA^-]}{[H^+]}$$

Collated:

$$[H^+]=\sqrt{\frac{K_{a1}(K_{a2}[HA^-]+K_S^{H_2O})}{K_{a1}+[HA^-]}} \tag{5-11a}$$

Formula (5-11a) is the exact formula for calculating the $[H^+]$ of HA^- amphoteric solution. Generally, the capacity of donating and accepting protons of HA^- is weak, then $[HA^-]\approx c$; when $K_{a2}\cdot c\geqslant 20K_S^{H_2O}$, water dissociation can be neglected; and if $c\geqslant 20K_{a1}$, $K_{a1}+c\approx c$, formula (5-18a) can be simplified as:

$$[H^+]=\sqrt{K_{a1}\cdot K_{a2}} \tag{5-11b}$$

Formula (5-11b) is the simplest formula for calculating the $[H^+]$ in HA^- amphoteric solution, which can be applied to other amphoteric substances.

For example, the simplest formula for calculating the $[H^+]$ in $H_2PO_4^-$ amphoteric solution is:

$$[H^+]=\sqrt{K_{a1}\cdot K_{a2}}$$

The simplest formula for calculating the $[H^+]$ in HPO_4^{2-} amphoteric solution is:

$$[H^+]=\sqrt{K_{a2}\cdot K_{a3}}$$

【Example 5-6】

Calculate the pH of 0.10mol/L $NaHCO_3$ aqueous solution.

Solution:

c=0.10mol/L, $K_{a1}=4.2\times10^{-7}$, $K_{a2}=5.6\times10^{-11}$

$\because K_{a2}\cdot c>20K_S^{H_2O}$, $c/K_{a1}>20$

$\therefore$ It can be solved by the application of the simplest formula:

$$[H^+]=\sqrt{K_{a1}\cdot K_{a2}}=\sqrt{4.2\times10^{-7}\times5.6\times10^{-11}}=4.8\times10^{-9}\text{mol/L}$$

$$\text{pH}=8.32$$

3.5 The calculation of $[H^+]$ in buffer solution

Buffer solution (缓冲溶液) is a kind of solution which can stabilize the acidity of solution. The buffer solution can resist changes in pH when acids or bases are added, or when a small amount of acids or bases produced by a chemical reaction, or when dilution occurs.

A buffer solution consists of a mixture of a weak acid and its conjugate base, or a weak base and its conjugate acid at predetermined concentrations or ratios, such as HAc-NaAc, NH_4Cl-NH_3. Here, we take a buffer solution composed of the monoprotic weak acid HA and its conjugate base NaA as an example to illustrate the calculation of $[H^+]$. The concentration of HA is c_{HA} and the concentration of its conjugate base NaA is c_{A^-}, respectively. Due to the presence of a large amount of HA and A^- in the solution at the same time, it is difficult to determine the zero level. Generally, the proton balance equation is derived from the mass balance equation and the charge balance equation.

Mass balance equation: $[HA]+[A^-]=c_{HA}+c_{A^-}$ $[Na^+]=c_{A^-}$

Charge balance equation: $[H^+]+[Na^+]=[OH^-]+[A^-]$

Collated:

$$[HA]=c_{HA}-[H^+]+[OH^-]$$

$$[A^-]=c_{A^-}+[H^+]-[OH^-]$$

From the dissociation equilibrium formula of [HA], we obtain:

$$[H^+]=K_a\frac{[HA]}{[A^-]}$$

Substituting the expressions of [HA] and $[A^-]$ into the above equation:

$$[H^+]=K_a\frac{c_{HA}-[H^+]+[OH^-]}{c_{A^-}+[H^+]-[OH^-]}$$

This is the exact formula for calculation of $[H^+]$ in buffer solution. In general, c_{HA}、$c_{A^-} \gg [H^+]$、$[OH^-]$, then $[H^+]$ and $[OH^-]$ can be ignored, so the above equation can be simplified as:

$$[H^+] = K_a \frac{c_{HA}}{c_{A^-}} \tag{5-12}$$

$$pH = pK_a + \lg \frac{c_{A^-}}{c_{HA}}$$

$[H^+] = K_a \dfrac{c_{HA}}{c_{A^-}}$ is the simplest formula for calculating the $[H^+]$ of buffer solution composed of a monoprotic weak acid and its conjugate base. Generally, as the concentration of buffer agent is large in the buffer solution that is used to control acidity, the calculation result is not required to be accurate, so the simplest formula can be used for calculation.

【Example 5-7】

(1) Add 50ml 0.10mol/L NaOH solution; (2) Add 50ml 0.10mol/L HCl solution to 200ml the buffer solution of 0.20mol/L NH_3-0.3mol/L NH_4Cl. How much has the pH of each solution changed? (The pK_a for NH_4^+ is 9.26)

Solution:

Calculate the pH of 0.20mol/L NH_3-0.3mol/L NH_4Cl buffer solution according to the simplest formula:

$$pH = pK_a + \lg \frac{c_{NH_3}}{c_{NH_4^+}} = 9.26 + \lg \frac{0.20}{0.30} = 9.08$$

At this point, $[H^+] = 10^{-9.08}$mol/L, because of c_{NH_3}、$c_{NH_4^+} \gg [H^+]$, $[OH^-]$ it is reasonable to use the simplest formula.

(1) Add 50ml 0.10mol/L NaOH solution:

$$c_{NH_3} = \frac{200\times0.20+50\times0.10}{200+50} = 0.18\text{mol/L}$$

$$c_{NH_4^+} = \frac{200\times0.30-50\times0.10}{200+50} = 0.22\text{mol/L}$$

Since both c_{NH_3} and $c_{NH_4^+}$ are large enough, the simplest formula is adopted for calculation:

$$pH = 9.26 + \lg \frac{0.18}{0.22} = 9.17$$

The pH of the solution increased by 9.17−9.08=0.09 pH units.

(2) Add 50ml 0.10mol/L HCl solution:

$$c_{NH_3} = \frac{200\times0.20-50\times0.10}{200+50} = 0.14\text{mol/L}$$

$$c_{NH_4^+} = \frac{200\times0.30+50\times0.10}{200+50} = 0.26\text{mol/L}$$

Apparently,

$$pH = 9.26 + \lg \frac{0.14}{0.26} = 8.99$$

The pH of the solution decreased by 9.08−8.99=0.09 pH units

Section 3 Acid-Base Indicators

1. The principle of discoloration of acid-base indicators

Acid-base indicators are a series of compounds that exhibit different colors corresponding to different values of hydrogen ion concentration or pH of the solution. They serve to signal the equivalence point. These indicators themselves are basically weak acids or weak bases which ionize as follows:

$$HIn \rightleftharpoons In^- + H^+$$

Where HIn is the acid form of the indicator showing one color (acid color) and In^- is the base form showing another color (base color).

2. The discoloration range of acid-base indicators and its influencing factors

For acid-base titrations, it is important to know the pH at which the indicator color change will occur. Assume the indicator is a weak acid, designated HIn, its dissociation equilibria is:

$$HIn \rightleftharpoons In^- + H^+$$

And its dissociation constant called indicator constant K_{HIn} is:

$$K_{HIn} = \frac{[H^+][In^-]}{[HIn]} \tag{5-13}$$

The above equation can be rearranged as:

$$\frac{[In^-]}{[HIn]} = \frac{K_{HIn}}{[H^+]}$$

The colors of the solution stained by the indicator is an average mixture of both acid colors and base colors, which responses directly to the concentration of acid forms and base forms, also H^+ in the solution. So the overall colors of an indicator can be decided by the ratio of $\frac{[In^-]}{[HIn]}$. However, the vision of the naked eyes of an analyst is not sensitive enough to identify the small color change. Eyes can generally discern only one color if it is 10 times as intense as the other is:

When the ionized form In^- is predominant, and $\frac{[In^-]}{[HIn]} \geqslant 10$, meanwhile $pH \geqslant pK_{HIn}+1$, the solution shows the basic colors.

When the nonionized form HIn is predominant, and $\frac{[In^-]}{[HIn]} \geqslant 10$, meanwhile $pH \geqslant pK_{HIn}-1$, the solution shows the acid colors.

An indicator should be chosen whose pK_{HIn} approximately equals the pH expected at the end point of the titration.

3. Common acid-base indicators

Phenolphthalein is a kind of organic weak acid indicators (K_a=6×10^{-10}), whose dissociation equilibria in an aqueous solution can be presented as follow:

Acid form (Colorless) $\xrightleftharpoons{pK_a = 9.1}$ Base form (Red) $+ H_2O$

Acid form
Colorless

Base form
Red

In an acid solution, phenolphthalein exists in the form of acid structure mainly and the solution shows colorless; when adding a base into this solution, the balance moves to the right and the acid form of phenolphthalein shifts to the base one, gradually. When pH≥10, phenolphthalein exists in the form of base structure mainly and the solution shows red. Conversely, phenolphthalein changes the color from red to colorless during adding acids.

Methyl orange is a kind of organic weak base indicators. In a base solution, methyl orange exists in the form of base structure mainly and the solution shows yellow; when adding an acid into this solution, the balance moves to the right and the base form of methyl orange shifts to the acid one. When pH≤3.1, methyl orange exists in the form of acid structure mainly and the solution shows red.

Base form (Yellow) $+ H^+ \xrightleftharpoons{pK_a = 3.45}$ Acid form (Red)

Base form
Yellow

Acid form
Red

Besides, some commonly used acid-base indicators are listed in Table 5-1.

Table 5-1 Common acid-base indicators

Indicators	pH range of color change	Color change	pK_{HIn}	Concentration	Dosage (drop/10ml agent)
Methyl orange	3.1–4.4	red → yellow	3.45	0.05% in aqueous solution	1
Bromophenol blue	3.0–4.6	yellow → purple	4.1	0.1% in 20% ethanol solution or its sodium brine solution	1
Bromocresol green	3.8–5.4	yellow → blue	4.9		1–3
Methyl red	4.4–6.2	red → yellow	5.0	0.1% in 60% ethanol solution or its sodium brine solution	1
Phenolphthalein	8.0–10.0	colorless → red	9.1	0.05% in 90% ethanol solution	1–3

4. Mixed acid-base indicators

As some acids and bases have a narrow pH range of titration jump, general acid-base indicators are not strong enough to detect these end points. Then, mixed acid-base indicators (混合酸碱指示剂) can be applied, which have the narrow range of color changes and the sharp color changes by using the complementary functions of colors.

There are two kinds of mixed acid-base indicators, mainly. The one is the mixture of an indicator and an inert dyestuff, and enhances the sharpness of color change based on the complementary functions of colors. For example, one mixed acid-base indicator consist of the methyl orange and the soluble indigo blue. The indigo does not change with pH change, which only contribute a blue background to methyl orange. When pH≥4.4, it shows green (combining yellow and blue); when pH=4.0, it shows light grey; When pH≤3.1, it shows purple (combining red and blue). The color change at the end point becomes very sharp.

The other one is the mixture of two or more than two indicators at a proper proportion, which has a narrower range of color change and a sharper color change based on the complementary functions of colors. For example, the transition range of bromocresol green (pK_{HIn}=4.9) is pH=3.8–5.4, and the transition range of methyl red (pK_{HIn}=5.0) is pH=4.4–6.2. When mixing bromocresol green and methyl red (3 : 1), due to the additive action of colors, the color change occurs at the pH of 5.1. The indicator's color change is wine red (the acid color)→light grey→green (the basic color). Whether the color change of the mixed acid-base indicator is significant or not, it depends on the property of indicators and dyestuff and the proportion of indicators. Table 5-2 gives several common mixed acid-base indicators.

Table 5-2 Common mixed acid-base indicators

Indicator Mixture	pH of Color Change	Color		Notes
		Acid Color	Basic Color	
0.1% methyl yellow ethanol solution +0.1% methylene blue ethanol solution (1∶1)	3.25	Bluish violet	Green	pH=3.4, green; pH=3.2, bluish violet
0.1% methyl orange aqueous solution +0.25% sodium indigotin-disulfonate aqueous solution (1∶1)	4.1	Purple	Olivine	pH=4.1, grey
0.1% bromocresol green ethanol solution +0.2% methyl red ethanol solution (3∶1)	5.1	Wine red	Green	
0.1% neutral red ethanol solution +0.1% methylene blue ethanol solution (1∶1)	7.0	Bluish violet	Green	pH=7.0, bluish violet
0.1% cresol red sodium salt aqueous solution +0.1% thymol blue sodium salt aqueous solution (1∶3)	8.3	Yellow	Green	pH=8.2, rose-bengal pH=8.4, purple

PPT

Section 4 Acid-Base Titration Curve and Indicator Selection

The mechanism of acid-base titration process can be understood by the change of hydrogen ion concentration during the course of the appropriate titration. pH change in the neighbourhood of the equivalence point called stoichiometric point is of the greatest importance, which provides the judgement that the tested substance can be titrated accurately or not (titration feasibility) and enables an indicator to be selected which will give a small titration error. The curve obtained by plotting pH as the ordinate against the percentage of acid neutralized or the number of ml of alkali added as abscissa is known as the titration curve (滴定曲线). To solve the above two fundamental problems of titration feasibility and indicator selection, the titration curves during three types of acid-base titration processes are discussed, then the titration feasibility and the indicator selection are determined according to the titration curve.

For each type of titration in this chapter, our goal is to construct a graph showing how pH changes with the titrant being added, which is known as theoretical titration curves. If you can do this, you will understand what is happening during the titration, and you will be able to interpret an experimental titration curve.

1. Strong acid (base) titration

As a simple example, let's focus on the titration of 20.00ml 0.1000mol/L HCl with 0.1000mol/L NaOH. The chemical reaction between the titrant and analyte is $H^+ + OH^- \rightleftharpoons H_2O$.

1.1 Titration curve

Three kinds of calculations must be done in order to construct the theoretical titration curve. Each of them corresponds to a distinct stage during the titration: (1) before the equivalence point, (2) at the equivalence point, and (3) After the equivalence point.

1.1.1 Before the equivalence point

Before the equivalence point, the pH is determined by the excess H^+ in the solution, we compute the concentration of the acid from its starting concentration and the amount of base added.

$$[H^+] = \frac{c_{HCl}V_{HCl} - c_{NaOH}V_{NaOH}}{V_{HCl} + V_{NaOH}}$$

At the beginning of the titration, $V_{NaOH} = 0$, $[H^+]$ in the solution is 0.1000mol/L, and the pH = 1.0.

When 19.98ml NaOH have been added, 99.9% of HCl have been titrated (0.1% of HCl have not been titrated, i.e., −0.1% relative error), the concentration of H^+ in the solution is

$$[H^+] = \frac{0.1000\times(20.00-19.98)}{20.00+19.98} = 5.0\times10^{-5}\text{mol/L}$$

$$pH = 4.3$$

1.1.2 At the equivalence point

At the equivalence point, the hydronium and hydroxide ions are present in equal concentrations, and the hydronium ion concentration is determined by the dissociation of water, and can be calculated directly from the ion-product constant for water, K_w, which is 1.0×10^{-14} at 25°C.

$$[H^+] = [OH^-]= \sqrt{K_w} = 1.0\times10^{-7}\text{mol/L}$$

$$pH = 7.0$$

1.1.3 After the equivalence point

After the equivalence point, the analytical concentration of the excess base is computed, and the hydroxide ion concentration is assumed to be equal to a multiple of the base concentration.

$$[OH^-] = \frac{c_{NaOH}V_{NaOH} - c_{HCl}V_{HCl}}{V_{NaOH} + V_{HCl}}$$

When 20.02ml NaOH have been added, excess 0.1% of NaOH solution have been added, the concentration of OH^- in the solution is

$$[OH^-]=\frac{0.1000\times20.02-0.1000\times20.00}{20.02+20.00}\approx 5.0\times10^{-5}\text{mol/L}$$

$$pOH=4.3 \qquad pH=9.7$$

The other pH values in the theoretical titration curves can be calculated according to the above methods, shown in Table 5-3. The titration curve obtained by plotting pH as the ordinate against the titration percentage of HCl neutralized or the number of ml of NaOH added as abscissa, is called the titration curve of a strong acid with a strong base (Figure 5-2). It is evident that the range from 20.00 to 40.00ml NaOH presents the reverse titration of the former 20.00ml NaOH in the presence of the non-hydrolysed sodium chloride solution.

Table 5-3 pH changes during the titration of 20.00ml 0.1000mol/L HCl with 0.1000mol/L NaOH (at room temperature)

Titration percentage (%)	Volume of NaOH added (ml)	Volume of excess HCl (ml)	$[H^+]$ (mol/L)	pH	
0	0.00	20.00	1.00×10^{-1}	1.00	
90.0	18.00	2.00	5.26×10^{-3}	2.28	
99.0	19.80	0.20	5.03×10^{-4}	3.30	
99.9	19.98	0.02	5.00×10^{-5}	4.30	pH jump range (99.9–100.1)
100	20.00	0.00	1.00×10^{-7}	7.00	
Titration percentage (%)	**Volume of NaOH added (ml)**	**Volume of excess NaOH (ml)**	**$[H^+]$ (mol/L)**	**pH**	
100.1	20.02	0.02	2.00×10^{-10}	9.70	
101	20.20	0.20	2.01×10^{-11}	10.70	
110	22.00	2.00	2.10×10^{-12}	11.68	
150	30.00	10.00	5.00×10^{-13}	12.30	
200	40.00	20.00	3.00×10^{-13}	12.52	

Characteristic of all useful titration is a sudden change in pH near the equivalence point. The data in Table 5-3 and the plot in Figure 5-2 show that as the titration proceeds, initially the pH rises slowly, the pH of the solution rises from 1.0 to 4.3 and only changes 3.3 pH units before the addition of 19.98ml NaOH, but it rises from 4.3 to 9.7 and changes 5.4 pH units between the addition of 99.9% and 100.1% of alkali and during the addition of 0.04ml alkali between 19.98ml and 20.02ml , i.e. the rate of pH change is

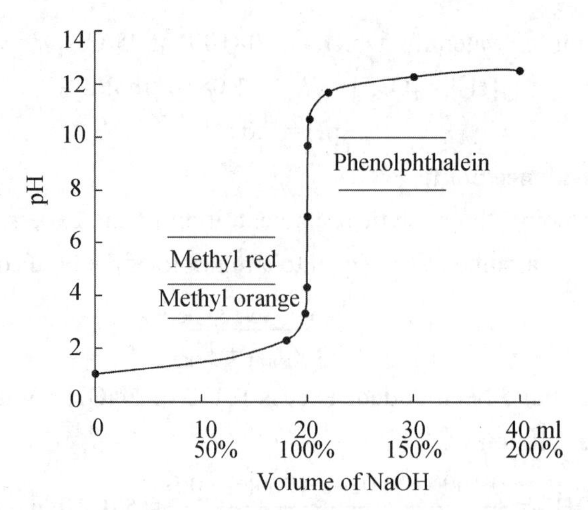

Figure 5-2 Titration curve for 20.00ml 0.1000mol/L HCl with 0.1000mol/L NaOH

very rapid in the vicinity of the equivalence point. After that, the rise of pH decreases gradually.

The rapid pH change in the solution near the equivalence point during the titration process is called as titration jump (滴定突跃) , and the pH range of titration jump is called as pH jump range (滴定突跃范围), i.e. the pH range between 0.1% relative error before and after the equivalence point. pH jump range is of the great importance, as it provides the judgement whether the tested substance can be titrated accurately or not (titration feasibility), and pH jump range is the basis for the indicator selection when the end point of titration is determined by the chemical indicator method.

1.2 Effect of titrant or analyte concentration on pH jump range

The effects of the titrant and analyte concentration on the neutralization titration curves for strong acids titrated are shown by the plots in Figure 5-3. The additions of bases have been extended in three different concentration of HCl, which concentration is 1.000mol/L, 0.1000mol/L and 0.0100mol/L HCl.

Note that with 0.1000mol/L NaOH as the titrant, the pH change near the equivalence point is large, and the pH jump range is 4.3–9.7. With 0.0100mol/L NaOH as the titrant, the pH change in the equivalence point region is much smaller, and the pH jump range is 5.3–8.7. On the contrary, with

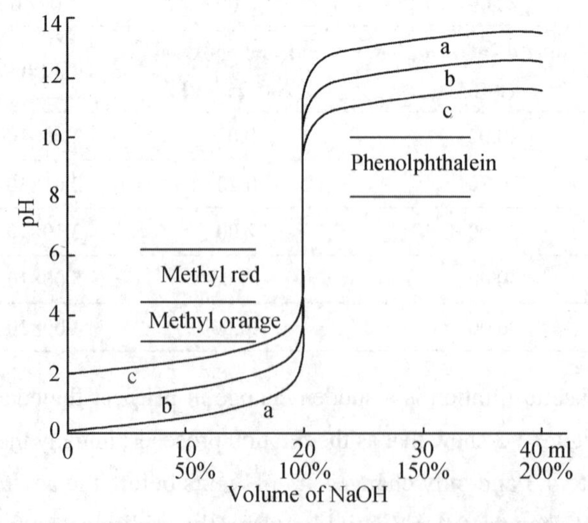

Figure 5-3 Titration curves for different concentration HCl with NaOH of equal concentration

(a) 1.000mol/L (b) 0.1000mol/L (c) 0.0100mol/L

1.000mol/L NaOH the titrant, the pH jump range is 3.3–10.7, much larger. Therefore, in the titration curve of a strong acid with a strong base, the pH jump range is only decided by the concentration of the acid and the base. For example, when the concentrations of the acid and the base both increase 10 times, the pH range of titration jump will enlarge 2 pH units.

1.3 Choosing an indicator

In quantitative analysis, the pH jump range is the basis for the indicator selection. We can conclude that if the end point of the titration is determined by some indicators which pH transition range falls into the pH jump range of the titration, the titration relative error will be smaller than 0.1% and the tested substance can be titrated accurately. For example, for the titration of 0.1000mol/L HCl solution with 0.1000mol/L NaOH solution, the pH jump range of titration is 4.3–9.7, and the titration end point can be determined by bromothymol blue (pH 6.2–7.6), phenol red (pH 6.8–8.4), methyl red (pH 4.4–6.2), methyl orange (pH 3.1–4.4), phenolphthalein (pH 8.0–10.0), etc. In that case, the volume differences for the titrations with the above indicators shown are of the same magnitude as the uncertainties associated with reading the buret and the titration errors negligible.

The pH jump range of the titration is wider, the indicator selection is more convenient, or the narrow pH jump range of titration limits the number of the usable indicators. For 1.000mol/L HCl titrated with 1.000mol/L NaOH, it is evident that any indicator with a transition range between 3.3 and 10.7 may be used, bromophenol blue and quinaldine red can be used except for the indicators selected in the titration of 0.1000mol/L HCl. For 0.01000mol/L HCl titrated, the ideal pH range is limited to 5.3–8.7, bromothymol blue, phenol red or methyl red will be suitable, but the titration error for methyl orange will be 1–2 percent, more than 0.1%.

Calculating the titration curve for the titration of a strong base with a strong acid is handled in the same manner, except that the strong base is in excess before the equivalence point and the strong acid is in excess after the equivalence point, and the titration curves for the acid and base titrated of equal concentration are linear symmetric, and the pH change directions of the above titration curves are inverse.

2. Unitary weak acid (base) titration

Unitary weak acid (base) titration includes the titration of a weak acid with a strong base and the titration of a weak base with a strong acid. In the titration of any weak acid (base), there are four regions of the titration curve that represent different kinds of calculation, i.e. (1) at the starting point of titration, (2) before the equivalence point, (3) at the equivalence point, (4) after the equivalence point.

(1) At the beginning, the solution contains only a weak acid or a weak base, and the pH is calculated from the concentration of the solute and its dissociation constant.

(2) After various volumes of the titrant have been added (not including the equivalence point), the solution consists of a series of buffers. The pH of each buffer can be calculated from the concentrations of the weak acid (base) that remains and its conjugate.

(3) At the equivalence point, the solution contains only the conjugate of the weak acid or base being titrated, and the pH is calculated from the concentration of this product.

(4) After the equivalence point, the excess of strong acid or base suppresses the acidic or basic character of the reaction product to some extent so that the pH is governed largely by the concentration of the excess titrant.

For this example, let's consider the titration of 20.00ml 0.1000mol/L acetic acid with 0.1000mol/L NaOH and plot the titration curve of a weak acid with a strong base.

2.1 Titration curve

2.1.1 At the starting point of titration

At the beginning, the solution contains only HAc and water, and the pH is calculated from the concentration of the solute and its dissociation constant in aqueous solution.

$$[H_3O^+] = \sqrt{c_a K_a}$$

c_a is the concentration of the weak acid, $c_{HAc} = 0.1000$mol/L;

K_a is the dissociation constant of the weak acid in aqueous solution, the dissociation constant of HAc in aqueous solution is 1.70×10^{-5}

$$[H_3O^+] = \sqrt{0.1000\times1.70\times10^{-5}} = 1.30\times10^{-3}\text{mol/L}$$

$$\text{pH} = -\lg(1.32\times10^{-3}) = 2.89$$

2.1.2 Before the equivalence point

From the first addition of the titrant until immediately before the equivalence point (not including the equivalence point), a portion of the acetic acid has been converted to its conjugate base, and the solution consists of a series of buffers. The pH of each buffer can be calculated from the concentrations of the weak acid, HAc, and its conjugate base, Ac^-. In Chapter 5.2, we can calculate the pH of a buffer using the Henderson–Hasselbalch equation.

$$\text{pH} = \text{p}K_a + \lg\frac{[Ac^-]}{[HAc]} \tag{5-14}$$

The equilibrium constant for the reaction between HAc and NaOH is large ($K=K_a/K_w=1.70\times10^9$), so we can treat the reaction as one that goes to completion. Before the equivalence point, the concentration of unreacted acetic acid is

$$[HAc] = \frac{c_a V_a - c_b V_b}{V_a + V_b}$$

and the concentration of acetate is

$$[Ac^-] = \frac{c_b V_b}{V_a + V_b}$$

For example, after adding 19.98ml NaOH and 99.9% of HAc titrated (−0.1% relative error), the pH of the solution is

$$\text{pH} = \text{p}K_a + \lg\frac{[Ac^-]}{[HAc]} = \text{p}K_a + \lg\frac{V_b}{V_a - V_b} = 4.76 + \lg\frac{19.98}{20.00-19.98} = 7.74$$

2.1.3 At the equivalence point

At the equivalence point, the moles of acetic acid initially present and the moles of NaOH added are identical. Since their reaction effectively proceeds to completion, the predominate ion in solution is Ac^-, which is a weak base. The pH is then calculated with the Formula (5-8) in Chapter 5.2 for a weak base. To calculate the pH, we first determine $[Ac^-]$.

$$[Ac^-] = \frac{c_b V_b}{V_a + V_b} = \frac{0.1000\times20.00}{20.00+20.00} = 0.05000\text{mol/L}$$

$$[OH^-] = \sqrt{c_b K_b} = \sqrt{0.5000\times\frac{1.0\times10^{-14}}{1.7\times10^{-5}}} = 5.42\times10^{-6}\text{mol/L}$$

$$pH = 14.00 - [-\lg(5.34\times10^{-6})] = 8.73$$

2.1.4 After the equivalence point

After the equivalence point, NaOH is present in excess, and the excess base and the acetate ion are both the sources of hydroxide ions. The contribution from the acetate ions is small and negligible because the excess of strong base suppresses the reaction between acetate and water. Then

$$[OH^-] = \frac{c_b V_b - c_a V_a}{V_a + V_b}$$

When 20.02ml NaOH have been added, 0.1% of excess NaOH have been added (+0.1% relative error), the concentration of OH^- in the solution is

$$[OH^-] = \frac{0.1000\times20.02 - 0.1000\times20.00}{20.02+20.00} \approx 5.0\times10^{-5}\text{mol/L}$$

$$pOH=4.30 \qquad pH=9.70$$

Table 5-4 show the calculated pH during this titration, and the calculated titration curve is shown in Figure 5-4. Characteristic of the titration curve for titrating 0.1000mol/L HAc with 0.1000mol/L NaOH is different from that for titrating 0.1000mol/L HCl with 0.1000mol/L NaOH.

Table 5-4 pH changes during the titration of 20.00ml 0.1000mol/L HAc with 0.1000mol/L NaOH (at room temperature)

Titration percentage (%)	NaOH added (ml)	Volume of excess HAc (ml)	Calculating equation	pH
0	0	20.00	$[H^+] = \sqrt{K_a \cdot c}$	2.88
50	10.00	10.00		4.76
90	18.00		$[H^+] = K_a \frac{[HAc]}{[Ac^-]}$	5.71
99.9	19.98			7.74
100	20.00		$[OH^-] = \sqrt{K_{b1} \cdot c}$	8.73 (pH jump range begins)
Titration percentage (%)	**NaOH added (ml)**	**Volume of excess NaOH (ml)**	**Calculating equation**	**pH**
100.1	20.02	0.02	$[OH^-] = 10^{-4.30}$	9.70 (pH jump range ends)
101	20.20	0.20	$[OH^-] = 10^{-3.30}$	10.70
110	22.00	2.00	$[OH^-] = 10^{-2.32}$	11.68
150	30.00	10.00	$[OH^-] = 10^{-1.70}$	12.30
200	40.00	20.00	$[OH^-] = 10^{-1.48}$	12.52

Firstly, the pH at the starting point of the titration curve for titrating HAc with NaOH is 2.88, and the pH at the starting point of the titration curve for titrating HCl with NaOH is 1.00, the former is about 2pH units higher than the latter.

Secondly, the section of the titration curve between the starting point and the equivalence point for titrating HAc with NaOH is also much higher than the latter. The solution consists of a series of buffers before the equivalence point and the pH of the solution rises slowly, and the section of the titration curve is flat. Near the equivalence point, there is a little HAc and a large amount of Ac^- in the solution, the dissociation effect of Ac^- in water increases, the buffer capacity of the solution decreases, the pH in solution and the slope of the titration curve increase rapidly.

Thirdly, because HAc is a weak acid, partially dissociated, there is a narrow pH jump range near the equivalence point in the titration curve, pH at the equivalence point is above 7.0 (pH 8.73) because Ac^- in aqueous solution appears alkaline, and the pH jump range is also in alkaline zone (pH range 7.74–9.70).

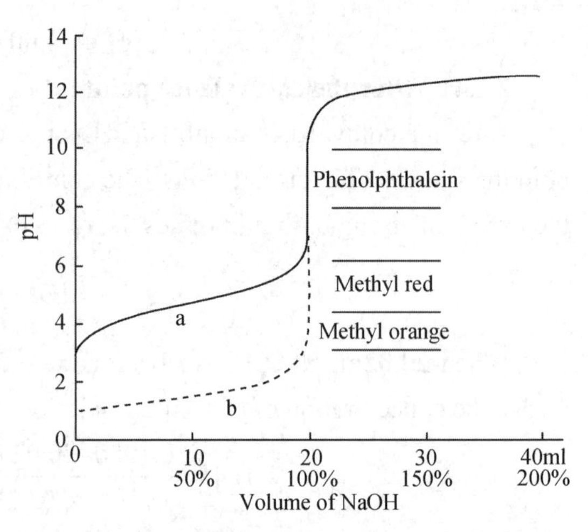

Figure 5-4 Titrative curve for 20.00ml 0.1000mol/L HAc (pK_a =4.76) with 0.1000mol/L NaOH

2.2 Indicator selection

For the titration of 0.1000 acetic acid with 0.1000 NaOH, the pH jump range is in the basic region (pH 7.74–9.70), and the titration end point should be determined by the indicators with a transition range on the alkaline side, such as phenolphthalein (pH 8.0–10.0) or thymol blue (pH 8.0–9.6). However, the indicators exhibiting a color change in the acidic region do not provide a sharp end point with a minimal error, such as methyl red, methyl orange, etc.

2.3 The effects of the acid concentration and strength on the titration curve

2.3.1 The effects of the acid concentration

Figure 5-5 is the titration curve of 0.1000mol/L and 0.001000mol/L HAc with NaOH of two equal concentrations. Note that the initial pH values are higher and the pH at stoichiometric point is lower for the more dilute solution (curve b in Figure 5-5). Before the equivalence point, however, the pH values differ only slightly because of the buffering action of the acetic acid/sodium acetate system present in this region, and the pH of the buffers is largely independent of the acid concentration. However, the pH jump range in the vicinity of the equivalence point becomes narrower with lower concentrations of the analyte and reagent, which is analogous to that for the titration of a strong acid with a strong base (Figure 5-3).

2.3.2 The effects of the acid strength

Titration curves for 0.1000mol/L solutions of acids with different dissociation constants are shown in Figure 5-6. Note that the starting point of the pH jump range is higher, and the pH jump range becomes narrower as the acid becomes weaker, i.e. dissociation constant of the acid is smaller, and the reaction between the acid and the base becomes less complete.

2.4 Titration feasibility

The pH jump range associated with titrating 0.001000mol/L HAc (curve b, Figure 5-5) is so narrow that there is likely to be a significant titration error regardless of any indicator. Figure 5-6 illustrates that similar problems occur as the strength of the acid titrated decreases. Therefore, if the dissociation constant of the acid is enough small and the concentration of the titrant is enough low, the weak acid cannot be titrated accurately. The pH jump ranges in titrating different strength acids with a strong base, i.e. 0.1% relative error before and after the stoichiometric point (sp) are listed in Table 5-5.

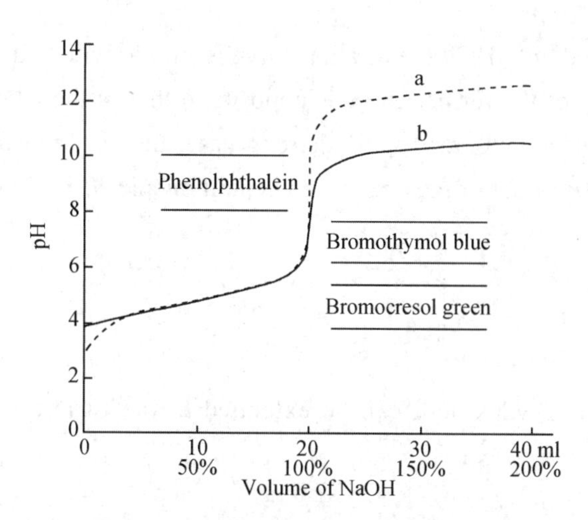

Figure 5-5 Titration curve of different concentration HAc with NaOH

(a) 0.1000mol/L HAc with 0.1000mol/L NaOH,
(b) 0.001000mol/L HAc with 0.001000mol/L NaOH.

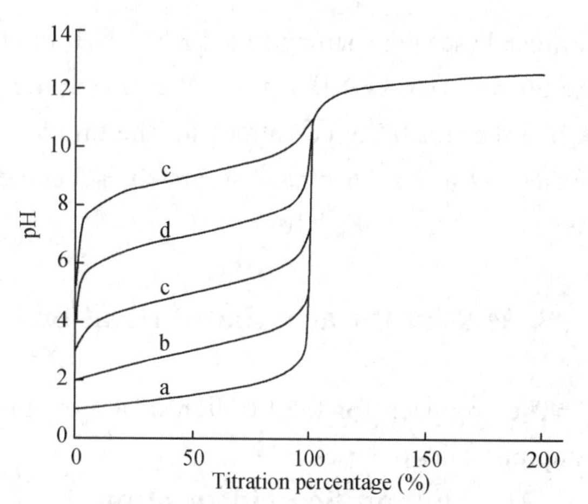

Figure 5-6 The effects of acid strength (dissociation constant) on titration curves

Each curve represents the titration of 20.00ml of 0.1000mol/L weak acid with 0.1000mol/L strong base.
(a) 10^{-5}, (b) 10^{-6}, (c) 10^{-7}, (d) 10^{-8}, (e) 10^{-9}

Table 5-5 pH jump range in titrating different strength acid with a strong base

K_a	pH jump range											
	1mol/L				0.1mol/L				0.01mol/L			
	−0.1%	sp	+0.1%	ΔpH	−0.1%	sp	+0.1%	ΔpH	−0.1%	sp	+0.1%	ΔpH
10^{-5}	8.00	9.35	10.70	2.70								
10^{-6}	9.00	9.85	10.70	1.70	9.00	9.35	9.77	0.77	8.70	8.85	9.00	**0.30**
10^{-7}	10.00	10.35	10.77	0.77	9.70	9.85	10.00	**0.30**	9.30	9.35	9.40	0.10
10^{-8}	10.70	10.85	11.00	**0.30**	10.30	10.35	10.40	0.10				
10^{-9}	11.30	11.35	11.40	0.10	10.83	10.85	10.87	0.04				

Fixed K_a and higher concentration, or fixed concentration and larger K_a, pH jump range (ΔpH) in the titration is bigger, i.e. if the product of the analyte concentration and K_a is larger, pH jump range is wider (Table 5-5). When the end point of the titration is judged accurately by human eyes with aid of an indicator with a transition color range, pH jump range (ΔpH in the titration have to be above 0.3pH unit, and the titration error will be below 0.1%. From the ΔpH in Table 5-5, accurate titration which end point is determined by an indicator requires that the product of the analyte concentration and K_a is above or equal to 10^{-8}. Therefore, the criterion for whether a weak acid is titrated accurately is usually the product of the analyte concentration and K_a is above or equal to 10^{-8}, i.e. $cK_a \geqslant 10^{-8}$.

As the pH jump range of titrating a weak acid with a strong base is in acidic region and its equivalence point is below 7.0, the titration end point should be determined by the indicators with a transition color range in acidic region, such as methyl red and methyl orange.

The calculations for the titration of a weak base with a strong acid are handled in a similar manner except that the initial pH is determined by the weak base, the pH at the equivalence point by its conjugate acid, and the pH after the equivalence point by the concentration of excess strong acid. For the titration

of a weak base with a strong acid, such as NH_3 titrated with HCl, the titration curve is similar with that of HAc titrated with NaOH, but the pH changing trend of the former curve is opposite to that of the latter. So, like the feasibility conditions for the titration of a weak acid with a strong base, the criterion for whether a weak base is titrated accurately is that the product of the analyte concentration and K_b is above or equal to 10^{-8}, i.e. $cK_b \geqslant 10^{-8}$.

3. Multiprotic acid (base) titration

The approach for the titration of a monoprotic weak acid can be extended to the titration of multiprotic acids or bases.

3.1 Multiprotic acid titration

Multiprotic acid dissociates stepwise in aqueous solution, for example the dissociation of H_2A in aqueous is as follow.

$$H_2A \rightleftharpoons H^+ + HA^-$$

$$K_{a1} = \frac{[H^+][HA^-]}{[H_2A]}$$

$$HA^- \rightleftharpoons H^+ + A^{2-}$$

$$K_{a2} = \frac{[H^+][A^{2-}]}{[HA^-]}$$

There are three principal questions to be solved in the titration of multiprotic acid. Firstly, whether $[H^+]$ from the gradual dissociation of multiprotic acid can be titrated accurately? Secondly, whether $[H^+]$ from the gradual dissociation of multiprotic acid can be titrated separately? Thirdly, If $[H^+]$ from the gradual dissociation of multiprotic acid can be titrated separately and accurately, how to select suitable indicators to determine the titration end point. The criteria for the above three questions are as follow.

(1) When $cK_{ai} \geqslant 10^{-8}$, there is a sharp pH jump near the equivalence point, and $[H^+]$ from this dissociation of multiprotic acid can be titrated accurately.

(2) When $K_{a1}/K_{a2} \geqslant 10^4$, i.e. the ratio between K_{a1} and K_{a2} is above and equal to 10^4, the two pH jumps near its two equivalence points can be separate, and $[H^+]$ from the gradual dissociation of multiprotic acid can be titrated separately, Or the two pH jumps near its equivalence points cannot be separate, and $[H^+]$ from the gradual dissociation of multiprotic acid can only be titrated together.

(3) Only the pH at the equivalence point of the titration is calculated for choosing an indicator conveniently, and the indicator with the transition color range near the above pH is selected for the titration end point.

pH values in the titration curve of multiprotic acid are usually calculated with the pH approximate calculated formula, because the less accuracy is required for the titration of multiprotic acid than that of monoprotic acid. For example, for 0.1000mol/L $H_2C_2O_4$ titrated with 0.1000mol/L NaOH, $c_1 \cdot K_{a1}=5.6\times10^{-2}\times0.1000>10^{-8}$, $c_2 \cdot K_{a2}=1.5\times10^{-4}\times0.1000/2>10^{-8}$, there are two pronounced pH jumps near the first equivalence point and the second equivalence point, and the first proton and the second proton dissociated from $H_2C_2O_4$ can both be titrated accurately. However, $\frac{K_{a1}}{K_{a2}} = \frac{5.6\times10^{-2}}{1.5\times10^{-4}} < 10^4$, the above two pH jumps near its two equivalence points cannot be separate, and the first proton and the second proton

ionized from $H_2C_2O_4$ cannot be titrated separately, and the total amount of two protons dissociated from $H_2C_2O_4$ are titrated together.

At the second equivalence point, the pH in the solution is determined from the dissociation of $C_2O_4^{2-}$, the concentration of hydroxide ion is

$$[OH^-]=\sqrt{c_b K_{b1}}=\sqrt{\frac{0.1000}{3}\times\frac{1.0\times10^{-14}}{1.5\times10^{-4}}}=1.5\times10^{-6}\,(mol/L)$$

$$pOH=5.82 \quad pH=8.18$$

Phenolphthalein is selected for the titration end point because its transition range is in the vicinity of the second equivalence point.

Triprotic acids (e.g., H_3PO_4) can be titrated similarly. When 0.1000mol/L H_3PO_4 titrated with 0.1000mol/L NaOH,

$$c_1 \cdot K_{a1}=6.9\times10^{-3}\times0.1000>10^{-8}$$

$$c_3 \cdot K_{a2}=6.2\times10^{-8}\times0.1000/2\approx10^{-8}$$

$$c_3 \cdot K_{a3}=4.8\times10^{-13}\times0.1000/3<10^{-8}$$

There is a pronounced pH jump near the first equivalence point or the second equivalence point, and the former two protons dissociated from H_3PO_4 can be titrated accurately, while the pH jump near the third equivalence point is not pronounced, the third proton dissociated from H_3PO_4 cannot be titrated accurately.

$$\frac{K_{a1}}{K_{a2}}=\frac{6.9\times10^{-3}}{6.2\times10^{-8}}>10^4,\quad \frac{K_{a2}}{K_{a3}}=\frac{6.2\times10^{-8}}{4.8\times10^{-13}}>10^4$$

It indicated that the two pH jumps near the first equivalence point and the second equivalence point are separate each other, the third proton dissociated from H_3PO_4 does not affect the accurate titration of the second proton dissociated from H_3PO_4. So the first proton and the second proton dissociated from H_3PO_4 can be titrated accurately and separately (Figure 5-7), the third proton dissociated from H_3PO_4 cannot be titrated with NaOH because the acidity of HPO_4^{2-} is weak and K_{a3} is below 10^{-7}.

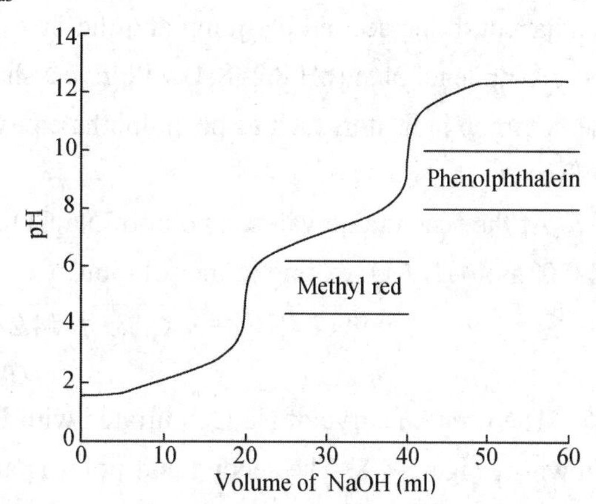

Figure 5-7 Titration curve of 0.1000mol/L H_3PO_4 with 0.1000mol/L NaOH

At the first equivalence point of H_3PO_4, a solution of $H_2PO_4^-$ exists, and methyl red and bromocresol green are the suitable indicators.

$$[H^+]\approx\sqrt{K_{a1}\cdot K_{a2}}$$

$$pH=\frac{1}{2}(pK_{a1}+pK_{a2})=4.69$$

At the second equivalence point of H_3PO_4, a solution of HPO_4^{2-} exists, phenolphthalein and thyme phthalein can be selected as the indicators.

$$[H^+]\approx\sqrt{K_{a2}\cdot K_{a3}}$$

$$pH=\frac{1}{2}(pK_{a2}+pK_{a3})=9.77$$

The titration error is relatively big when the above common indicators are used because the common

indicators do not give sharp color changes while the pH jump ranges are narrow. If the mixed indicators instead of the above common indicator, such as the mixed indicator of bromocresol green-methyl orange instead of bromocresol green or methyl orange, the mixed indicator of phenolphthalein-thyme phthalein instead of phenolphthalein or thyme phthalein, the titration error will decrease.

3.2 Multiprotic base titration

Similar with multiprotic acid, multiprotic base also dissociates stepwise in aqueous solution, the answer for whether $[OH^-]$ from the gradual dissociation of multiprotic base can be titrated separately or accurately can be judged referring to the criteria for the titration of multiprotic acid, and the titration end point is determined by an indicator with a transition range near the equivalence point.

For example, Na_2CO_3 is a dibasic base, $K_{b1}=1.8\times10^{-4}$, $K_{b2}=2.4\times10^{-8}$. For 0.1000mol/L Na_2CO_3 titrated with 0.1000mol/L HCl, $c_1\cdot K_{b1}=1.8\times10^{-4}>10^{-8}$, $c_2\cdot K_{b2}=2.4\times10^{-9}\approx10^{-8}$, there is respectively a pronounced pH jump in the vicinity of the first equivalence point or the second equivalence point, and OH^- dissociated from CO_3^{2-} and HCO_3^- can both be titrated accurately by HCl. $K_{b1}/K_{b2}\approx10^4$, and the pH jump near the first equivalence point and the second equivalence point are separate, and OH^- dissociated from CO_3^{2-} and HCO_3^- can be titrated separately by HCl.

At the first equivalence point of Na_2CO_3 titrated with HCl, a solution of HCO_3^- exits,

$$[H^+]\approx\sqrt{K_{a1}\cdot K_{a2}}=\sqrt{4.2\times10^{-7}\times5.6\times10^{-11}}=4.8\times10^{-9}\ (mol/L)$$

$$pH=8.31$$

Phenolphthalein and thyme blue can be selected as the indicators for the first equivalence point of Na_2CO_3. As HCO_3^- has a relatively strong buffer capacity at the first equivalence point, the first pH jump is adjacent to the second pH jump and the first pH jump range is narrow, the mixed indicators of cresol red-bromphenol blue (pH 8.2–8.4) will give a sharper end point near the first equivalence point instead of the common indicator, such as phenolphthalein, which color change is not obvious in the titration of base with acid.

At the second equivalence point of Na_2CO_3, the pH of the solution is determined by the dissociated of 0.03mol/L H_2CO_3 exiting in the solution.

$$[H^+]=\sqrt{K_{a1}\cdot c}=\sqrt{4.2\times10^{-7}\times0.03}=1.1\times10^{-4}\ (mol/L)$$

$$pH=3.96$$

The titration curve of Na_2CO_3 titrated with HCl is shown in Figure 5-8. The second end point is always used for the quantitative titration because the second pH jump range is wider than the first. Methyl orange or bromphenol blue is usually selected for the second end point of Na_2CO_3, but neither of them gives a sharp end point because K_{b2} is still small and the pH jump range near the second equivalence point is narrow. The mixed indicator of methyl orange-bromocresol green is used instead. To prevent the end point from advancing because of the generation of CO_2 supersaturated solution, shake the solution violently near the second equivalence point, or boil the solution to decompose H_2CO_3 into CO_2 and expel CO_2, cool the solution and continuously titrate the solution after it is cold. So a sharper end point can be obtained by

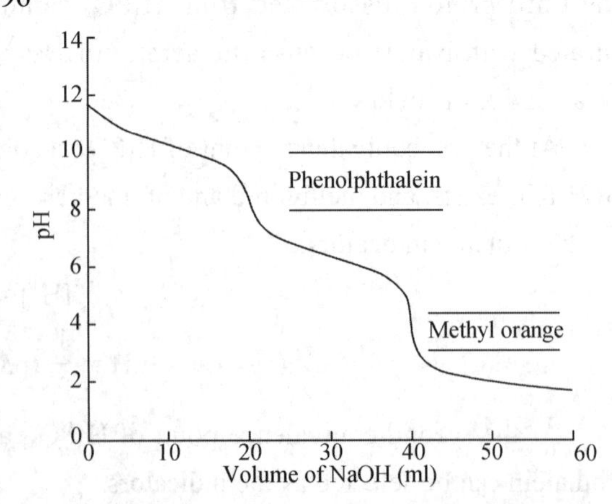

Figure 5-8 Titration curve of 0.1000mol/L Na_2CO_3 with 0.1000mol/L HCl

boiling the solution briefly to eliminate the reaction product, H_2CO_3 or CO_2.

4. Titration error

There is an important distinction between an end point and an equivalence point or a stoichiometric point. The difference between these two terms is important. The equivalence point or stoichiometric point occurs when stoichiometrically equal amounts of analyte and titrant react. An equivalence point, therefore, is a theoretical not an experimental value. An end point for a titration is determined experimentally and represents the analyst's suitable estimate of the corresponding equivalence point. In the acid-base titration, the end point is usually monitored by a visual indicator, and gets close to the equivalence point, but not at the equivalence point. Any difference between the equivalence point and the end point of a titration is a principal source of titration error, the relative error caused by their difference is called titration error (滴定误差), which is expressed as TE%. Titration error is a kind of method error, which is determined by the amount of the excess acid(base) titrated or the excess titrant.

$$\text{TE\%} = \frac{\text{the amount of the insufficient titrant or excess titrant}}{\text{the amount of the analyte}} \times 100\%$$

4.1 Titration error of a strong acid (base)

Assumed that the initial concentration of the analyte is c_0, the initial volume of the analyte is V_0, the volume of the titrant titrated at the end point is V.

The pH of solution is equal to 7.0 at the equivalence point of a strong acid(base) titrated with a strong base(acid). If the end point coincides with the equivalence point, the pH of solution is equal to 7.0 at the end point, i.e. $[H^+]_{ep}=[OH^-]_{ep}$, and TE%=0. $[H^+]_{ep}$ or $[OH^-]_{ep}$ is respectively the concentration of hydrogen ion or hydroxyl ion in the solution at the end point.

For the titration of a strong acid with a strong base, if the end point is before the equivalence point and the pH of solution is below 7.0, i.e. $[H^+]_{ep}>[OH^-]_{ep}$, there is a small amount of excess acid in the solution, and the titrant insufficient, TE% < 0, negative error exists. H^+ in the solution at the end point comes from the dissociation of the excess acid and water. If the dissociation of water is negligible, the concentration of excess acid in the solution at the end point is $c_{a\,excess}=[H^+]_{excess}=[H^+]_{ep}-[OH^-]_{ep}$.

$$\text{TE\%} = -\frac{[H^+]_{excess}}{c_{ep}} \times 100\% = \frac{[OH^-]_{ep}-[H^+]_{ep}}{c_{ep}} \times 100\% \qquad (5\text{-}14a)$$

c_{ep} is the analytical concentration of the analyte in the solution at the end point, $c_{ep}=\frac{c_0V_0}{V_0+V}$, the volume of the titrant is usually approximate to the initial volume of the analyte, i.e. $V=V_0$, $c = 1/2\ c_0$.

If the end point is after the equivalence point and the pH of solution is above 7.0, i.e. $[H^+]_{ep} < [OH^-]_{ep}$, there are a small amount of excess titrant in the solution, TE%>0, positive error exists. OH^- in the solution at the end point comes from the dissociation of the excess base and water. If the dissociation of water is negligible, the concentration of excess base is $c_{b\,excess}=[OH^-]_{excess}=[OH^-]_{ep}-[H^+]_{ep}$.

$$\text{TE\%} = \frac{[OH^-]_{excess}}{c_{ep}} \times 100\% = \frac{[OH^-]_{ep}-[H^+]_{ep}}{c_{ep}} \times 100\% \qquad (5\text{-}14b)$$

Similarly, the calculating formula of titration error for titrating a strong base is formula (5-15).

$$TE\% = \frac{[H^+]_{ep} - [OH^-]_{ep}}{c_{ep}} \times 100\% \quad (5\text{-}15)$$

【Example 5-8】

In the titration of 20.00ml 0.1000mol/L HCl with 0.1000mol/L NaOH, if the end point is determined by methyl orange (pH=4.0) or phenolphthalein (pH=9.0), Calculate the titration error of them respectively.

Solution:

When the end point is determined by methyl orange, pH=4.0, $[H^+]_{ep}=1.0\times10^{-4}$mol/L, $[OH^-]_{ep}=1.0\times10^{-10}$mol/L, c_{ep}=0.1000/2=0.05000mol/L

$$TE\% = \frac{[OH^-]_{ep} - [H^+]_{ep}}{c_{ep}} \times 100\% = \frac{1.0\times10^{-10} - 1.0\times10^{-4}}{0.05000} \times 100\% = -0.2\%$$

When the end point is determined by phenolphthalein, pH=9.0, $[H^+]_{ep}= 1.0\times10^{-9}$mol/L, $[OH^-]_{ep} = 1.0\times10^{-5}$mol/L, c_{ep}= 0.1000/2 =0.05000mol/L

$$TE\% = \frac{[OH^-]_{ep} - [H^+]_{ep}}{c_{ep}} \times 100\% = \frac{1.0\times10^{-5} - 1.0\times10^{-9}}{0.05000} \times 100\% = 0.02\%$$

4.2 Titration error of unitary weak acid (base) titration

For the titration of a weak acid with a strong base, if the end point is before the equivalence point, there are an amount of excess weak acid in the solution, and the titrant insufficient, TE% < 0, and negative error exists. The proton generated in the solution comes from the partly dissociation of the excess weak acid and water, while the strong titrant and water consume the proton in the titration process. According to the proton balance equation, i.e. the amount of the protons generated is equal to the amount of the protons consumed in the titration, the concentration of excess acid in the solution at the end point is $c_{a\ excess} = [OH^-]_{ep} - ([H^+]_{ep} + [HA]_{ep})$.

$$TE\% = \frac{[OH^-]_{ep} - ([H^+]_{ep} + [HA]_{ep})}{c_{ep}} \times 100\% \quad (5\text{-}16)$$

$[HA]_{ep}$, $[H^+]_{ep}$, $[OH^-]_{ep}$ are the concentration of HA, H^+, OH^- in the solution at the end point, c_{ep} is the analytical concentration of the analyte in the solution at the end point, $c_{ep} = \frac{c_0 V_0}{V_0 + V} = \frac{1}{2} c_0$.

At the end point of a weak acid titrated with a strong base, the solution is alkaline, $[H^+]_{ep} \ll [OH^-]_{ep}$, and $[H^+]_{ep}$ in the solution may be usually negligible, and Formula (5-16) is simplified as Formula (5-17).

$$TE\% = \frac{[OH^-]_{ep} - [HA]_{ep}}{c_{ep}} \times 100\% \quad (5\text{-}17)$$

$$[HA]_{ep} = \delta_{HA} \cdot c_{ep} = \frac{[H^+]_{ep}}{[H^+]_{ep} + K_a} \cdot c_{ep}$$

So the titration error of a weak acid can be simplified as Formula (5-18).

$$TE\% = \left(\frac{[OH^-]_{ep}}{c_{ep}} - \frac{[H^+]_{ep}}{[H^+]_{ep} + K_a} \right) \times 100\% \quad (5\text{-}18)$$

The calculating formula of titration error for the titration of a weak base with a strong acid is as follow.

$$TE\% = \frac{[H^+]_{ep} - ([OH^-]_{ep} + [A^-]_{ep})}{c_{ep}} \times 100\% \quad (5\text{-}19)$$

In the titration of a weak base with a strong acid, the solution is acidic at the end point, $[H^+]_{ep} \gg$

$[OH^-]_{ep}$, and $[OH^-]_{ep}$ in the solution may be negligible, and Formula (5-19) is simplified and Formula (5-20) and Formula (5-21).

$$TE\% = \frac{[H^+]_{ep} - [A^-]_{ep}}{c_{ep}} \times 100\% \tag{5-20}$$

$$[A^-]_{ep} = \delta_{A^-} \cdot c_{ep} = \frac{K_a}{[H^+]_{ep} + K_a} \cdot c_{ep}$$

$$TE\% = \left(\frac{[H^+]_{ep}}{c_{ep}} - \frac{K_a}{[H^+]_{ep} + K_a}\right) \times 100\% \tag{5-21}$$

【Example 5-9】

In the titration of 20.00ml 0.1000mol/L HAc with 0.1000mol/L NaOH, if the end point is determined by phenolphthalein (pH=9.0), Calculate the titration error.

Solution:

When the end point is determined by phenolphthalein, pH=9.0, $[H^+]_{ep}$= 1.0×10^{-9}mol/L, $[OH^-]_{ep}$ = 1.0×10^{-5}mol/L, K_a =1.7×10^{-5}, c_{ep}= 0.1000/2 =0.05000mol/L

$$TE\% = \left(\frac{1.0\times10^{-5}}{0.05000} - \frac{1.0\times10^{-9}}{1.0\times10^{-9} + 1.7\times10^{-5}}\right) \times 100\% = 0.014\%$$

Section 5 Applications of Acid-Base Titrations

PPT

Acid-base titrations are widely used to determine the concentration of acidic or basic analytes or analytes that can be converted into acids or bases by appropriate treatments. Water is the common solvent for neutralization titrations because it is convenient, inexpensive, nontoxic and low expansion coefficient. However, some analytes cannot be titrated in aqueous media because of their low solubility or their weak acidity or basicity not enough to give satisfactory end points, the above substances can often be titrated in nonaqueous solvents. We shall restrict our discussions to aqueous systems in Chapter 5.5.

1. Preparation and calibration of standard acid or alkaline solutions

We noted that the reaction between strong acids and strong bases is more complete, and produce the larger pH jump range near the equivalence point. For this reason, standard solutions for acid-base titration are always prepared from strong acids or strong bases, and HCl (NaOH) is the most frequently-used standard acid (alkaline) solution, and the concentration of standard solutions are usually 0.1mol/L, 1.0mol/L or 0.01mol/L.

Most standard acid or alkaline solutions are prepared by the calibration method, such as HCl solution and NaOH solution. Firstly, a solution of an approximate concentration is prepared by diluting the concentrated reagent to accurately known volumes. the diluted solution is then standardized against a primary-standard-grade reagent or another standard solution.

1.1 Standard 0.1mol/L HCl

1.1.1 Preparation of 0.1mol/L HCl

The molar concentration of ~37% wt% HCl from commercial sources is about 12mol/L. To obtain 1000ml 0.1mol/L HCl by diluting the concentrated HCl, the calculated volume of the concentrated HCl is 8.3ml. In practice, the volume of the concentrated HCl to be diluted is a little more than the calculated volume, and 9.0ml concentrated HCl is taken and diluted to 1000ml, then the diluted acid solution is then standardized against a primary-standard base.

1.1.2 Calibration of 0.1mol/L HCl

Sodium carbonate or sodium tetraborate decahydrate is the frequently used reagent for standardizing acids.

Primary-standard-grade sodium carbonate is available commercially or can be prepared by heating purified sodium hydrogen carbonate between 270 to 300°C until a constant weight. Because Na_2CO_3 absorbs moisture in the air easily, dry pure Na_2CO_3 for 1 h at 110°C before use and store it in a closed container, cool the container in a desiccator, weigh Na_2CO_3 rapidly as required to avoid the error from the moisture in the air. The calibration reaction between Na_2CO_3 and HCl is

$$Na_2CO_3 + 2HCl \rightleftharpoons 2NaCl + H_2O + CO_2\uparrow$$

The product at the second equivalence point is H_2CO_3 (pH=3.89), and the mixed indicator of methyl orange-bromocresol green is usually used to determine the end point, but methyl orange cannot give a sharp end point. The accurate concentration of HCl is calculated according to the volume of HCl added.

$$c_{HCl} = \frac{2m_{Na_2CO_3}}{V_{HCl} \cdot M_{Na_2CO_3}} \times 1000$$

The molar mass of Na_2CO_3 is small, which leads to a relatively big error when weighed. In addition, the moisture absorbed by Na_2CO_3 is another error source.

The molar mass of sodium tetraborate decahydrate ($Na_2B_4O_7 \cdot 10H_2O$) is relatively big, and $Na_2B_4O_7 \cdot 10H_2O$ does not absorb moisture and primary-standard-grade $Na_2B_4O_7 \cdot 10H_2O$ is easily got. However, $Na_2B_4O_7 \cdot 10H_2O$ loses its crystal water when the relative humidity in the air is below 39%. Primary-standard-grade $Na_2B_4O_7 \cdot 10H_2O$ should be reserved in the closed container with the relative humidity of 60%. The reaction between sodium tetraborate decahydrate and HCl is

$$Na_2B_4O_7 + 2HCl + 5H_2O \rightleftharpoons 4H_3BO_3 + 2NaCl$$

The extremely weak acid of H_3BO_3 (K_a =5.4×10^{-10}) is produced and pH of the solution is 5.1 at the equivalence point, the common indicator of methyl red gives a sharp end point.

$$c_{HCl} = \frac{2m_{Na_2B_4O_7 \cdot 10H_2O}}{V_{HCl} \cdot M_{Na_2B_4O_7 \cdot 10H_2O}} \times 1000$$

In addition, the concentration of HCl can be calibrated with a standard NaOH solution with an accurate concentration, and phenolphthalein is used for the end point.

1.2 Standard 0.1mol/L NaOH

1.2.1 Preparation of 0.1mol/L NaOH

NaOH is the most used base for preparing standard solutions. NaOH cannot be obtained in primary-standard purity, because it absorbs moisture and CO_2 in the air easily, and so, standard NaOH solution need to be standardized after its preparation.

Absorption of CO_2 by a standardized solution of sodium hydroxide leads to a negative systematic error in the analyses in which an indicator with a basic discoloration range is used; there is no systematic error when an indicator with an acidic discoloration range is used, i.e pH at end point is below 6.0.

To obtain 1000ml 0.1mol/L NaOH solution without CO_3^{2-}, prepare saturated aqueous NaOH solution (52 wt%, the density is close to 1.56g of solution per milliliter) which molar concetration is 20mol/L in advance and allow the Na_2CO_3 precipitate to settle overnight. (Na_2CO_3 is insoluble in this solution.) Store the solution in a tightly sealed polyethylene bottle and avoid disturbing the precipitate when supernate is taken. The calculated volume of the supernate is 5.0ml, and 5.6ml supernate is taken and diluted to 1000ml. Note that new boiled and cooled distilled water without CO_2 is used when supernate is diluted.

1.2.2 Calibration of 0.1mol/L NaOH

Potassium hydrogen phthalate (KHP) or oxalic acid ($H_2C_2O_4 \cdot 2H_2O$) is the frequently used reagent for standardizing base, and standard HCl solution with accurate concentration is also used.

KHP can be obtained in primary-standard purity after recrystallized and used to calibrate the basic standard solution. The molar mass of KHP (204.2g/mol) is relatively big, and it does not absorb moisture and has no crystal water, and so, it is easy to be preserved. KHP is the most widely used primary standard substance. Dry potassium hydrogen phthalate for 1 h at 110°C and store it in a desiccator before used.

$$KHC_8H_4O_4 + NaOH \rightleftharpoons KNaC_8H_4O_4 + H_2O$$

The secondary dissociation constant of KHP(K_{a2}) is 3.9×10^{-6}, and pH of the solution at the equivalence point is 9.1, the common indicator of phenolphthalein gives a sharp end point. The concentration of NaOH is calculated with the volume of NaOH added.

$$c_{NaOH} = \frac{m_{KHP}}{V_{NaOH} \cdot M_{KHP}} \times 1000$$

Oxalic acid ($H_2C_2O_4 \cdot 2H_2O$) in primary-standard purity is stable relatively, it will not lose its crystal water when be reserved in the closed container with the relative humidity of 5%-95%. The reaction between oxalic acid and NaOH is

$$H_2C_2O_4 + 2NaOH \rightleftharpoons Na_2C_2O_4 + 2H_2O$$

The secondary dissociation constant of $H_2C_2O_4 \cdot 2H_2O$ (K_{a2}) is 1.5×10^{-4}, and pH of the solution at the equivalence point is 8.4, the common indicator of phenolphthalein is used. The concentration of NaOH is calculated with the volume of NaOH added.

$$c_{NaOH} = \frac{2m_{H_2C_2O_4 \cdot 2H_2O}}{V_{NaOH} \cdot M_{H_2C_2O_4 \cdot 2H_2O}} \times 1000$$

2. Application examples

Acid-base titrations are used to determine the amount of many inorganic, organic, and biological species that possess acidic or basic properties, such as alkaloid, aromatic acid, etc. In addition, the analytes such as esters or lactones, which can be converted into acids or bases by suitable chemical treatments, can be titrated with a standard strong base or acid.

2.1 Direct titration

If $c \cdot K_a \geqslant 10^{-8}$ or $c \cdot K_b \geqslant 10^{-8}$, the compounds with acidic or basic properties can be titrated directly with the strong standard solution. $c \cdot K_{ai} \geqslant 10^{-8}$ and $\frac{K_{a_i}}{K_{a_{i+1}}} \geqslant 10^4$ or $\frac{K_{b_i}}{K_{b_{i+1}}} \geqslant 10^4$ are the criteria for whether multiprotic acid (base) or the mixtures of acids or bases can be titrated accurately and separately, or the above acid(base) or their mixtures need to be titrated indirectly.

2.1.1 Determination of the content of total alkaloids in Aconiti Radix

Alkaloids are the bioactive compounds in Aconiti Radix, such as aconitine, hypaconitine and mesaconitine, etc. The content of total alkaloids in Aconiti Radix can be titrated directly by strong standard acids.

Procedure: weigh accurately 5.0mg alkaloid extract in Aconiti Radix, add 5ml of neutral ethanol, slightly heat and dissolve the extract, then add 30ml of cool new boiled distilled water and 4 drops of 0.1% methyl red, titrate the solution with 0.02000mol/L HCl untill the color of the solution turns into red. The amount of total alkaloids is calculated by aconitine, 1ml 0.02000mol/L HCl represents 0.0129g aconitine. The content of total alkaloids is calculated with the volume of HCl titrated and the weight of the extract.

$$\text{Total alkaloids\%} = \frac{T_{T/A}}{m_S} V_T \times 100\%$$

$T_{T/A}$ is the titer of aconitine by 0.02000mol/L HCl, V_T is the volume of HCl standard solution titrated, m_S is the accurate mass of the extract.

2.1.2 Titration the mixtures of NaOH and Na_2CO_3 (double indicator titration)

The sample of NaOH, sometimes containing a little amount of Na_2CO_3 because it absorbs CO_2 in the air during its storage or transport, can be titrated with a strong acid to determine the amount of each base in the solution. The content of NaOH or Na_2CO_3 is determined separately by double indicator titration. One using an alkaline discoloration indicator, such as phenolphthalein, and the other with an acidic discoloration indicator, such as bromocresol green or methyl orange. The volumes of strong acid needed to titrate OH^- to the pH 8.3 end point and the pH 3.9 end point will be the same, while titrating CO_3^{2-} to the pH 3.9 end point requires twice as much strong acid as titrating it to the pH 8.3 end point requires.

$$NaOH + HCl \rightleftharpoons NaCl + H_2O \qquad (pH=7.0)$$

$$Na_2CO_3 + HCl \rightleftharpoons NaHCO_3 + NaCl \qquad (pH=8.3)$$

$$NaHCO_3 + HCl \rightleftharpoons NaCl + H_2O + CO_2\uparrow \qquad (pH=3.9)$$

Firstly, add phenolphthalein into the analyte's solution, titrate the mixtures with a standard HCl to the phenolphthalein end point, i.e. the color of the solution changes from red to colorless. At this point, NaOH has been titrated wholly, $NaHCO_3$ has been produced from Na_2CO_3, and the volume of HCl added is V_1 ml. Then, add methyl orange into the analyte solution, titrate the solution continuously until methyl orange turns into orange red, $NaHCO_3$ has been titrtated wholly, and the volume of HCl consumed later is V_2 ml. For the mixtures of NaOH and Na_2CO_3, the volume of standard HCl reacting with Na_2CO_3 is $2V_2$ ml, and the volume of standard HCl reacting with NaOH is (V_1-V_2) ml, the contents of NaOH and Na_2CO_3 are calculated according to the following equations, in which S is the mass of the sample weighed, M_{NaOH} or $M_{Na_2CO_3}$ is the malar mass of NaOH or Na_2CO_3 respectively.

$$NaOH\% = \frac{c_{HCl}(V_1 - V_2) \cdot M_{NaOH}}{m_S \times 1000} \times 100\%$$

$$Na_2CO_3\% = \frac{c_{HCl} V_2 \cdot M_{Na_2CO_3}}{m_S \times 1000} \times 100\%$$

The method described above is not entirely satisfactory because the pH change at the hydrogen carbonate end point (pH~8.3) is not sufficient to give a sharp color change with a chemical indicator (Figure 5-8). Because of this lack of sharpness at this end point, relative errors of 1% or more are common. A suitable acidic discoloration indicator such as the mixed indicator of methyl orange-bromocresol green is selected at the second end point and this indicator will give a sharp color change by expelling CO_2 in the

solution. The total alkaline content of the mixtures of NaOH and Na_2CO_3 can be calculated by Na_2O, and the calculating formula of $Na_2O\%$ is as follow.

$$Na_2O\% = \frac{c_{HCl} \cdot (V_1 + V_2) \cdot M_{Na_2O}}{2m_S \times 1000} \times 100\%$$

On the other hand, double indicator titration can determine qualitatively the compositions of the mixtures containing Na_2CO_3, $NaHCO_3$, and NaOH or either alone. The compositions of the alkaline solution can be deduced from the relationships of V_1 and V_2 (Table 5-6).

Table 5-6 Volume Relationships in the analysis of the mixtures containing Na_2CO_3, $NaHCO_3$, and NaOH

Relationships of V_1 and V_2	Compositions of the alkaline mixtures
$V_1 \neq 0$, $V_2 = 0$	NaOH
$V_1 = 0$, $V_2 \neq 0$	$NaHCO_3$
$V_1 = V_2 \neq 0$	Na_2CO_3
$V_1 > V_2 > 0$	NaOH and Na_2CO_3
$V_2 > V_1 > 0$	$NaHCO_3$ and Na_2CO_3

Once the composition of the alkaline solution has been established, the volume of standard NaOH solution can be used to determine the concentration of each component in the sample.

2.2 Indirect titration

The dissociation constants in aqueous solution for most of organic acid and organic base in traditional Chinese medicine preparation are between 10^{-6}-10^{-9}, in which the weak acids ($c \cdot K_a \leqslant 10^{-8}$) or bases ($c \cdot K_b \leqslant 10^{-8}$) or the analytes insoluble in water cannot be titrated directly in aqueous solution. However, the analytes insoluble in water and the above weak acid (base) can be titrated in non-aqueous solution, or those weak acids (bases) can be back-titrated or indirect-titrated after their acidity (alkalinity) are strengthened in aqueous solution, for example boric acid or organic nitrogen.

2.2.1 Titration of boric acid

Boric acid (K_a=5.4×10^{-10}) is too weak to be directly titrated with strong bases. H_3BO_3 can react with polybasic alcohols, such as glycol or glycerol, to form complex acids. The dissociation constants of these complex acids are about 10^{-4}–10^{-7} and these complex acids can be titrated with standard NaOH solution. The complex acid of H_3BO_3 and glycerol (K_a =3.0×10^{-7}) is as follow.

$$2\ \text{[OH, OH, OH]} + H_3BO_3 \rightleftharpoons \left[\text{(O, O)B(O, O); OH, HO}\right]^- + H^+ + 3H_2O$$

$$\left[\text{(O, O)B(O, O); OH, HO}\right]^- + H^+ + NaOH \rightleftharpoons \left[\text{(O, O)B(O, O); OH, HO}\right]^- + Na^+ + H_2O$$

Procedure: weigh accurately 0.2g H_3BO_3 (dried in a desiccator with H_2SO_4 in advance), add 30ml mixtures of distilled water and glycerol (1 : 2), slightly heat and dissolve H_3BO_3, cool the solution to the room temperature rapidly, add 3 drops phenolphthalein, then titrate the solution with 0.1mol/L standard NaOH until the color of the solution turns into pink. The content of H_3BO_3 is calculated according to the following equation.

$$H_3BO_3\% = \frac{c_{NaOH} \cdot V_{NaOH} \cdot M_{H_3BO_3}}{m_S \times 1000} \times 100\%$$

2.2.2 Determination of the content of nitrogen in organic compounds (Kjeldahl nitrogen analysis)

Kjeldahl analysis is used frequently to determine the nitrogen content in organic substances containing proteins, amino acids and alkaloids, etc. In the Kjeldahl method, a sample is firstly decomposed in boiling sulfuric acid, which converts amine and amide nitrogen into ammonium ion, organic substance into CO_2 and water. $CuSO_4$ or mercury salts is added as a catalyst to speed the reaction, the boiling point of concentrated (98 wt%) sulfuric acid (338℃) is raised by adding K_2SO_4. The above process is called as digestion, and digestion is carried out in a long-neck Kjeldahl flask that prevents loss of sample from spattering. After digestion is complete, the solution containing NH_4^+ is cooled down and a concentrated NaOH aqueous solution is added carefully which neutralizes the H_2SO_4 in the solution and evolves ammonia from ammonia salt. The liberated NH_3 is steam distilled (with a large excess of steam) into a receiver containing saturated H_3BO_3 solution, then $H_2BO_3^-$ produced in the solution is titrated with standard H_2SO_4 to determine the content of nitrogen in the sample.

Kjeldahl digestion　　organic C,H,N $\xrightarrow{H_2SO_4}$ $NH_4^+ + CO_2 + H_2O$

Neutralization of NH_4^+　　$NH_4^+ + OH^- \rightleftharpoons NH_3(g) + H_2O$

NH_3 distilled and absorbed　　$H_3BO_3 + NH_3 \rightleftharpoons NH_4^+ + H_2BO_3^-$

Titration　　$2H_2BO_3^- + H_2SO_4 \rightleftharpoons 2H_3BO_3 + SO_4^{2-}$

The protein in dry yeast is more than 40% and contains mainly glycine, the content of nitrogen in dry yeast may be determined by the Kjeldahl method and then the content of protein can be calculated from a knowledge of the nitrogen percentage contained in it.

Procedure: weigh accurately 0.2g dry yeast on filter paper, put the sample weighed and filter paper together into a 500ml Kjeldahl flask. Added 5g anhydrous Na_2SO_4 and 0.3g anhydrous $CuSO_4$ successively, and then add slowly 10ml concentrated H_2SO_4 along the flask wall. Tilt the Kjeldahl flask at 45 degree angle, heat the flask slowly with direct fire and keep the solution not boiling until boiling bubble stops, then heat the solution rapidly until boiling, continue heating the solution so that the solution boils gently for 30 minutes after it becomes transparent green, and then cool the solution. Add 125ml water along the flask wall slowly, shake and mix the solution, add 60ml 40% NaOH along the flask wall after the solution turns cold. Add several Zinc particles and connect the Kjeldahl flask and the condenser tube with nitrogen balloon, dip the other end of the condenser tube just into 50ml 2% H_3BO_3 which is put into a 500ml conical flask, shake the Kjeldahl flask slightly and mix the solution, and then distill the solution by heating until the received solution in the conical flask is up to 250ml. Lower the receiver and rinse the condenser with a little water. Add 4 drops the mixed indicators of methyl red-bromocresol green and titrate the distillate with 0.05mol/L H_2SO_4 until the color of the solution turns into gray-purple. Repeat a blank titration and correct the determination result. 1ml 0.05mol/L H_2SO_4 represents 1.401mg nitrogen.

$$N\% = \frac{2c_{H_2SO_4} \cdot V_{H_2SO_4} \cdot M_N}{m_S \times 1000} \times 100\% \quad \text{or} \quad N\% = \frac{T_{H_2SO_4/N} \cdot V_{H_2SO_4}}{m_S \times 1000} \times 100\%$$

The content of nitrogen in different protein are almost the same, the conversion factor of nitrogen and protein is 6.25, in the other word, the content of nitrogen in proteins is about 16%. So the content of protein in dry yeast is calculated according the following equation. If most of the proteins in the sample are albumin, the conversion factor of nitrogen and protein is 6.27.

$$\text{Protein\%} = \frac{\dfrac{c_{H_2SO_4}}{0.05} \cdot V_{H_2SO_4} \times 1.401 \times 6.25}{m_S \times 1000} \times 100\%$$

The organic substances containing nitro groups, nitroso groups and azo groups need to be reduced before they are digested, and then the groups containing nitrogen are converted quantitatively into NH_4^+. Ferrous salt, thiosulfate and glucose are the mostly used reagents for reducing the oxidizing nitrogen.

Section 6 Acid-Base Titrations in Nonaqueous Solutions

PPT

1. Basic principles of nonaqueous acid-base titrations

Although aqueous solution is wildly used in titration analysis, there are still many titrations difficult to perform in aqueous medium due to several reasons: the substance has a poor solubility in water, or the dissociation equilibrium constant K of weak acids or bases is too low to meet 10^{-7} for an accurate titration. While failures of titrations in aqueous solution are met for such reasons, here are chances that they can be rescued by the use of different nonaqueous solvents, like organic solvents. Vorlander was the first to use a titration in nonaqueous medium. In 1903, he titrated aniline with hydrochloric acid dissolved in benzene. The application of nonaqueous media in the titration of organic compounds is expanding, especially in the pharmaceutic field.

1.1 Classification of solvents

The practical classification for solvents, consistent with “acid-base titrations”, is classifying them briefly as acidic, basic, or neutral. Another more insightful system involves merely two kinds, protonic solvents and aprotic solvents, based on whether the solvent is able to auto-dissociate. These two classifications often work together to estimate the character of a certain solvent.

Aprotic solvents do not exhibit much detectable acid or base properties, or react with any other acid or base compounds dissolved. Among which, the most inert ones are benzene and alkanes (chloroform, carbon tetrachloride). Others can show slight proton-acceptor activity, such as ketones; or with little basic properties because of weakly basic oxygen or nitrogen conducts interaction with acids through hydrogen bonds, such as pyridine.

1.2 Properties of solvents

The main determinants of the property of a solvent discussed below involve: autoprotolysis constant, acidity or basicity, dielectric constant and polarity, leveling effect and differentiate effect.

1.2.1 Autoprotolysis constant of solvents

In an amphiprotic solvent,

$$HS \rightleftharpoons S^- + H^+$$
$$HS + H^+ \rightleftharpoons H_2S^+$$

In the equation, HS is the undissociated form of a general solvent, H_2S^+ is the lyonium ion protonated

from HS, and S^- is the lysing ion deprotonated from HS.

The autoprotolysis reaction is obtained by adding the two reactions above:

$$2HS \rightleftharpoons S^- + H_2S^+$$

And the equilibrium constant of the above reaction is:

$$K^{HS} = \frac{[H_2S^+][S^-]}{[HS]^2} = K_a^{HS} K_b^{HS}$$

$$K_s^{HS} = [H_2S^+][S^-] = K_a^{HS} K_b^{HS} [HS]^2 \quad (5\text{-}22)$$

[HS] is approximate to a constant because of weak self-ionization of the solvent, so K_s^{HS} is a constant at a certain temperature and called the autoprotolysis constant of solvents. For example, the autoprotolysis constant of H_2O, $K_s^{H_2O} = [H_3O^+][OH^-] = 1.0\times10^{-14}$ (25℃). Other dissociative solvents are similar to water, the autoprotolysis reaction of ethanol is

$$2C_2H_5OH \rightleftharpoons C_2H_5OH_2^+ + C_2H_5O^-$$

$$K_s^{C_2H_5OH} = [C_2H_5OH_2^+][C_2H_5O^-] = 7.9\times10^{-20} \text{ (25℃)}$$

At a certain temperature, different solvents have different autoprotolysis constants because of their different dissociation. The autoprotolysis constant of common solvents are listed in Table 5-7.

Table 5-7 The autoprotolysis constant of common solvents (25℃)

Solvents	Methanol	Formic acid	Acetic acid	Acetic anhydride	Ethylenediamine	Acetonitrile
K_s^{HS}	2×10^{-17}	6×10^{-7}	3.6×10^{-15}	3×10^{-15}	5×10^{-16}	3×10^{-27}

The degree of the acid-base reaction is closely related to the dissociation of solvents, and the degree of the same acid-base reaction in a solvent with a small autoprotolysis constant is more complete than that in another solvent with a big autoprotolysis constant. The equilibrium constant of a strong acid-base reaction in a solvent HS is K_t^{HS}.

$$K_t^{HS} = \frac{1}{K_s^{HS}}$$

For example, the equilibrium constant of a strong acid-base reaction in water at 25℃ is 1.0×10^{14}, while the equilibrium constant of a strong acid-base reaction in ethanol at 25℃ is 1.2×10^{19}, so the degree of the same acid-base reaction in ethanol is more complete than that in water, and pH jump range of the strong acid-base titration in ethanol is wider, which is shown in Table 5-8.

Table 5-8 pH jump range of 0.1mol/L HCl in water and ethanol solvent

Solvents	pK_s	sp−0.1%	sp+0.1%	pH jump range
H_2O	14.00	pH=4.3	pH=14.00−4.3=9.7	4.3–9.7
C_2H_5OH	19.10	$pCH_3CH_2OH_2^+$=4.3	$pCH_3CH_2OH_2^+$=19.10−4.3=14.8	4.3–14.8

1.2.2 Acidity of HA and basicity of B

Acidity of HA or basicity of B can be measured by the acid dissociation constant K_a or the base dissociation constant K_b, while K_a^{HA} and K_a^{B} present their inherent acidity and inherent basicity.

$$HA \rightleftharpoons A^- + H^+$$

$$K_a^{HA} = \frac{[H^+][A^-]}{[HA]} \quad (K_a^{HA}\text{, Inherent acidity constant})$$

$$B + H^+ \rightleftharpoons HB^+$$

$$K_b^B = \frac{[HB^+]}{[H^+][B]} \quad (K_b^B\text{, Inherent basicity constant})$$

When HA dissolved in protonic solvents, i.e. HS, the reaction between HA and HS undergoes two steps: the first is ionization by the proton exchange with the solvent, the second is a further dissociation of cations and anions by the dispersal of solvent molecules.

$$HA + HS \rightleftharpoons (H_2S^+ \cdot A^-) \rightleftharpoons H_2S^+ + A^-$$

The equilibrium constant of the above reaction is

$$K_{a(HA)} = \frac{[H_2S^+][A^-]}{[HS][HA]} = K_a^{HA} K_b^{HS} \tag{5-23}$$

The acidity of HA in HS solvent is determined by the inherent basicity constant of HS and the inherent acidity constant of HA, i.e. proton-donor ability of HA and proton-acceptor of HS.

Similarly, B dissolved in HS solvent, the reaction between B and HS is

$$B + HS \rightleftharpoons S^- + HB^+$$

The equilibrium constant of the above reaction is

$$K_{b(B)} = \frac{[HB^+][S^-]}{[B][HS]} = K_b^B K_a^{HS} \tag{5-24}$$

The basicity of B in HS solvent is determined by the inherent basicity constant of B and the inherent acidity constant of HS, i.e. proton-donor ability of Hs and proton-acceptor of B.

For example, the inherent acidity constant of , $K_a^{HA} = 1.0\times10^{-3}$, dissolved in HS ($K_b^{HS} = 1.0\times10^{-3}$) and HS* ($K_b^{HS} = 1.0\times10^{3}$) with different inherent basicity and same dielectric constants, the acidity of HA in HS and HS* is respectively as follow.

In HS, $K_{a(HA)} = K_a^{HA} K_b^{HS} = 1.0\times10^{-3}\times10^{-3} = 1.0\times10^{-6}$, HA is a weak acid.

In HS*, $K_{a(HA)} = K_a^{HA} K_b^{HS} = 1.0\times10^{-3}\times10^{3} = 1.0$, HA is a strong acid.

Therefore, acidity of HA or basicity of B is related to not only their own inherent acidity or basicity but also the inherent basicity or acidity of solvents.

1.2.3 Dielectric constant and polarity of solvents

Dielectric constant (represented by D), also called permittivity, is an important physical quantity for the choice of a solvent (Table 5-9).

The electric energy E of an ion is given by:

$$E = \frac{\varepsilon^2 z + z}{2Dr}$$

Where ε is the unit charge, z is the valency of the ion, D is the dielectric constant of the solvent, and r is the radius of the ion in solution. Known from the equation, the dielectric constant is inversely proportional to the potential. The dielectric constant of the solvent is proportional to polarity, a more polar solvent shows larger dielectric constant (D_{water}=78.5), while a less polar solvent shows smaller dielectric constant ($D_{benzene}$=2.285). In solvent with low dielectric constant, the Coulomb attraction between product ions will affect the dissociation of acids and bases, so they are hard to dissociate further into free ions and exist in ion pairs, with their acidity or basicity decreased compared to in solvent with high dielectric constant.

Table 5-9 Dielectric constants of common solvents

Solvent	Dielectric constant	Temperature (°C)
Cyclohexane	2.03	15
1,4-Dioxane	2.24	20
Benzene	2.28	20
Propanoic acid	3.44	40
Diethyl ether	4.24	20
Chloroform	4.81	20
Acetic acid	6.15	20
Pyridine	12.01	20
Methyl isobutyl ketone	13.11	20
Ethylenediamine	16.00	18
2-Butanone	18.98	15
1-Butanol	19.20	30
Acetic anhydride	20.50	20
Acetone	20.70	20
Ethanol	24.30	25
Methanol	32.63	25
Acetonitrile	36.00	20
Ethanolamine	37.70	20
Dimethylformamide	38.30	20
Formic acid	58.50	15

Leveling effect and differentiating effect of solvents

For example, four mineral acids: perchloric acid, sulfuric acid, hydrochloric acid, and nitric acid, are all strong acids dissociate completely in the water:

$$HClO_4 + H_2O \rightleftharpoons H_3O^+ + ClO_4^-$$
$$H_2SO_4 + H_2O \rightleftharpoons H_3O^+ + HSO_4^-$$
$$HCl + H_2O \rightleftharpoons H_3O^+ + Cl^-$$
$$HNO_3 + H_2O \rightleftharpoons H_3O^+ + NO_3^-$$

Actually, they do not have same intrinsic acid strengths. But in water, their dissociation equilibria all lie extremely to the right, so their apparent acid strengths are all finally reduced to the same level of H_3O^+, as there is no acid stronger than H_3O^+ existing in water. We name this phenomenon as the leveling effect (拉平效应), as well as water a leveling solvent (拉平溶剂) for the mineral acids.

However, in acetic acid solution, their acid strengths can be distinct:

$$HClO_4 + HAc \rightleftharpoons H_2Ac^+ + ClO_4^-$$
$$H_2SO_4 + HAc \rightleftharpoons H_2Ac^+ + HSO_4^-$$
$$HCl + HAc \rightleftharpoons H_2Ac^+ + Cl^-$$
$$HNO_3 + HAc \rightleftharpoons H_2Ac^+ + NO_3^-$$

Acetic acid is less basic than water and does not level strong acids. The measurement of the equilibrium constants shows that, from the top to the bottom, the reaction is getting less complete. So, we call this opposite phenomenon differentiating effects (区分效应), and acetic acid a differentiating solvent (区分溶剂) for the four mineral acids.

It should be noticed that the definition of a leveling or differentiating solvent depends on the species being titrated. For bases ammonia, diethylamine, and triethylamine, they are strong enough to react sufficiently with acetic acid, acetic acid hereby acts as a leveling solvent. While $K_a^{H_2O}$ is smaller than K_a^{HAc}, the reaction between different base and H_2O are not complete according to the Formula (5-24), they are of different base strength in water, thus water here is a differentiating solvent.

If the inherent acidity or basicity of solvents are similar, the levelling effect is stronger for the solvent with a bigger dielectric constant, the differentiating effect is strong for the solvent with a small dielectric constant. For example, D_{HAc}=6.15 and D_{HCOOH}=58.5, the differentiating effect between $HClO_4$, H_2SO_4, HCl of HAc is stronger than that of formic acid.

Autoprotolysis constant of HS(K_s^{HS}) also affects the leveling effect and differentiating effect of solvents. When K_s^{HS} is small, acids or bases of different strength may be differentiated in this solvent, on the contrary, acids or bases of different strength cannot be coexisted in the solvent.

Therefore, the leveling effect and differentiating effect are relative. In general, acidic solvents are the leveling solvents for bases and the differentiating solvents for acids, basic solvents are the leveling solvents for acids and the leveling solvents for bases. Aprotic solvents are often used to differentiate the mixed acids or base. For example, ketones are aprotic and neutral solvents in general use. Among which, methyl isobutyl ketone is especially good at the titration of a mixture of acids. The graph on the left shows the simultaneous titration of five different acids with base (0.2mol/L tetrabutylammonium hydroxide) in methyl isobutyl ketone. The end points separate clearly by the order of acid strength ($HClO_4$ strongest, phenol weakest) with insignificant leveling effects.

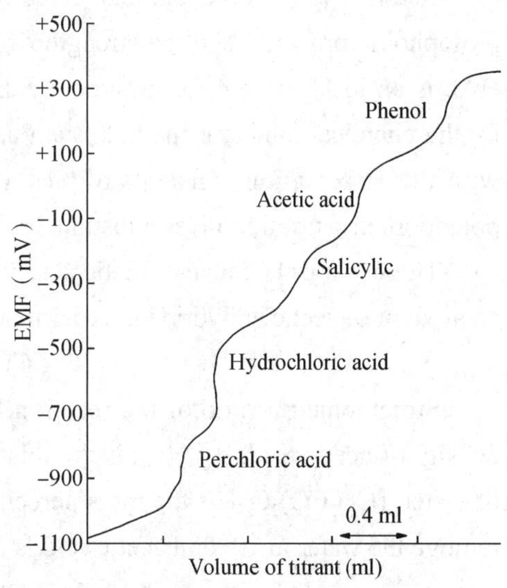

Figure 5-9 Separate titration curves of five mixed acids

1.3 Selection of solvents

The selection of solvents is critical before any nonaqueous titration is performed. For the given solvent used as the titration medium, a number of requirements must be met:

(1) The solvent should have suitable acid/base property, to favor the acid-base reaction go rapid and complete. Generally speaking, weak bases will be more sufficient titrated in acidic solvents, likewise weak acids will be more sufficient titrated in basic solvents.

(2) The solvent should have a low dielectric constant, which can make the end point sharper and more available to detect.

(3) The solvent can solve the sample and its products.

(4) The solvent requires high purity, low and volatility.

(5) Mixed solvents can be used to make the sample dissolved and the end point sharper. For polar samples, a solvent with strong polarity is selected to dissolve the sample, then an appropriate solvent

with weak polarity is added to decrease the dielectric constant and sharpen the end point. So the common mixed solvents are usually composed of protonic solvents and aprotic solvents according to a certain proportion, such as diols and hydrocarbons, HAc-benzene or HAc-CCl_4, or benzene-mathanol, or benzene-pridine, or benzene-isopropanol, etc.

2. Titration of acids and bases in non-aqueous solutions

2.1 Titration of bases

2.1.1 Non-aqueous solvents

Usually acidic solvents should be selected to increase the strength of weak bases for titration, such as acetic acid, formic acid, methylacetic acid and nitromethane, etc. For some insoluble samples or some titrations without a distinct end point, the mixed solvents of protonic solvents and aprotic solvents are selected to dissolve the samples or give the titration a sharp end point, For example, the titration in the mixed solvent of HAc-CCl_4 will be given a sharper end point than that in HAc solvent.

Acetic acid is stable and unaffected by the air. HAc is more acidic than water because of its strong protophobic property, and the strength of a weak base increases if it placed in HAc solvent. Therefore, HAc is an ideal solvent for titrating weak bases with the dissociation constants of base (K_b) over 10^{-11} by the chemical indicator method, such as crystal violet as the indicator. For the extremely weak bases with the dissociation constants of base (K_b) below 10^{-12}, neither the chemical indicator titration nor potentiometric titration gives a distinct end point.

There is a little water in acetic acid from commercial sources, and the water need to be removed by an amount of acetic anhydride in order to avoid affecting the titration.

$$(CH_3CO)_2O + H_2O \rightleftharpoons 2CH_3COOH$$

Stoichiometric ratio of the reaction between $(CH_3CO)_2O$ and H_2O is 1∶1. d_{CH_3COOH} is the relative density of acetic acid, $d_{(CH_3CO)_2O}$ is the relative density of acetic anhydride, $H_2O\%$ is the mass percentage of water, $(CH_3CO)_2O\%$ is the mass percentage of acetic anhydride, the volume of $(CH_3CO)_2O$ needed to remove the water in 1000ml acetic acid is calculated with the following equation.

$$\frac{d_{CH_3COOH}\times 1000\times H_2O\%}{M_{H_2O}}=\frac{d_{(CH_3CO)_2O}\times V_{(CH_3CO)_2O}\times[(CH_3CO)_2O]\%}{M_{(CH_3CO)_2O}}$$

$$V_{(CH_3CO)_2O}=\frac{d_{CH_3COOH}\times 1000\times H_2O\%\times M_{(CH_3CO)_2O}}{d_{(CH_3CO)_2O}\times M_{H_2O}\times[(CH_3CO)_2O]\%}$$

For example, if $H_2O\%$ in acetic acid is 0.20% and its relative density is 1.05, the relative density of 97.0% (wt %) $(CH_3CO)_2O$ is 1.08, the volume of $(CH_3CO)_2O$ needed to remove the water in 1000ml acetic acid is calculated as follow.

$$M_{H_2O}=18.02\text{g/mol},\quad M_{(CH_3CO)_2O}=102.09\text{g/mol}$$

2.1.2 Standard solution and primary standard substances

Acetic acid can differentiate many inorganic acids as a solvent, the strength of the acids in HAc solvent decreases in the following order, which is $HClO_4>HBr>H_2SO_4>HCl>HNO_3$. $HClO_4$ is the strongest acid in acetic acid solvent and stable, and the products from organic or inorganic base titrated with $HClO_4$ are easily dissolved in acetic acid solvent, so $HClO_4$ in acetic acid solvent is widely used as a standard solution to titrate the base in non-aqueous solution.

(1) Preparation of 0.1mol/L $HClO_4$-HAc standard solution. The $HClO_4$ reagent from commercial

sources is an aqueous solution containing 70.0%–72.0% (wt%) $HClO_4$, and an amount of acetic anhydride need to be added to remove the water in the $HClO_4$ reagent. 8ml 72% (wt%) $HClO_4$ with the relative density of 1.75 is taken to prepare 1000ml 0.1mol/L $HClO_4$-HAc solution, and in order to remove the water in the above 8ml $HClO_4$ reagent, the volume of 97.0% (wt%) $(CH_3CO)_2O$ with the relative density of 1.08 is 20.20ml.

Notes that 72% (wt%) $HClO_4$ cannot be exposed to the evaluated temperature or organic substance because they may be explosive easily, and cannot be directly mixed with $(CH_3CO)_2O$ because of the dangers caused by a lot of heat released in this process. At first, 72% (wt%) $HClO_4$ should be diluted with HAc, then a proper quantity of $(CH_3CO)_2O$ is added slowly while the solution is stirred so that the temperature of the solution is below 25°C.

If the samples are easy for acetylation, the water content in the sample need to be determined by Karl Fisher titrimetry and adjusted to 0.01%–0.2% with water and $(CH_3CO)_2O$, or excess $(CH_3CO)_2O$ makes the determined result lower. Usually the amount of $(CH_3CO)_2O$ used is slightly more than the calculated amount in the titration of other common samples, which will not affect the determined result.

The mixed solvent of HAc- $(CH_3CO)_2O$ (9 : 1) is adopted to prepare the $HClO_4$ non-aqueous solution as the $HClO_4$-HAc solution will freeze if the temperature is below 16°C and its use will be affected. Sometimes 10%–15% methylacetic acid is added into acetic acid to prevent the solution frozen.

(2) Calibration of 0.1mol/L $HClO_4$-HAc standard solution. Potassium hydrogen phthalate (KHP) in primary-standard purity is usually used to calibrate the concentration of 0.1mol/L $HClO_4$-HAc standard solution, and crystal violet is used as the indicator, the reaction between KHP and $HClO_4$ is as follow.

(3) Temperature calibration of $HClO_4$-HAc standard solution. The concentration of acid or base aqueous solution is not affected by the storage temperature because the expansion coefficient of water (2.1×10^{-4}/°C) is small. The expansion coefficients of most organic solvents are relatively big, for example, the expansion coefficient of acetic acid (1.1×10^{-3}/°C) is 5 times of that of water, and the volume of acetic acid has 0.11% relative error when the temperature has changes 1°C. Therefore, the concentration of $HClO_4$-HAc standard solution need to be recalibrated when the temperature change is over 10°C compared with the temperature of the solution being calibrated. If the above temperature change is below 10°C, the concentration of $HClO_4$-HAc standard solution can be calibrated with the following equation. a is the expansion coefficient of HAc, t_0 is the temperature of the solution being calibrated, t_1 is the temperature of the solution to determine the content of the analyte, C_0 is the original molar concentration of $HClO_4$ in $HClO_4$-HAc standard solution, C_1 is the molar concentration of $HClO_4$ recalibrated in the standard solution.

$$C_1 = \frac{C_0}{1+a(t_1-t_0)}$$

2.1.3 Indicators

The most frequently used indicator in the titration of base with $HClO_4$-HAc standard solution is crystal violet. Crystal violet is a multiprotic base and can accept several protons. The acidic color of crystal violet is purple, and its basic color is yellow, blue-purple, blue, blue-green, yellow-green are the transitional colors of crystal violet. The dissociation equilibrium of crystal violet in the solution of different pH is as follow.

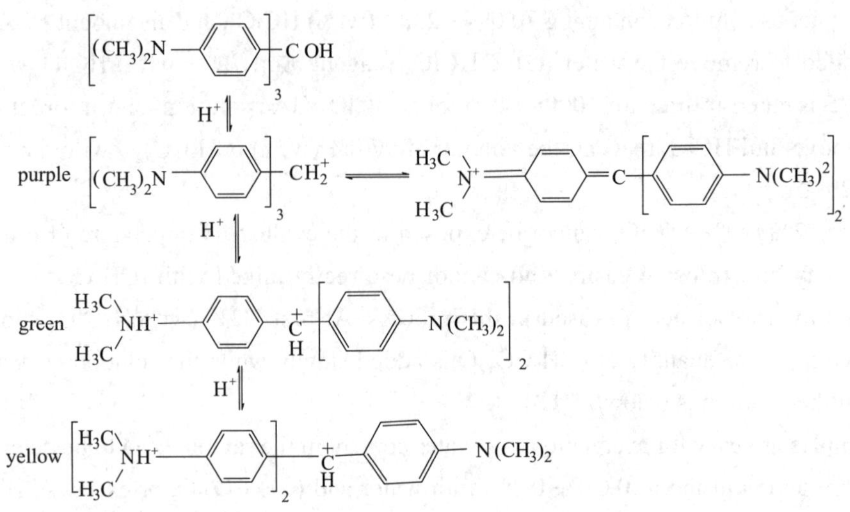

purple

green

yellow

Crystal violet gives different color end-points in the titration of different strength base, the color of end point for the stronger base is blue or blue-green, and the color of end point for the weaker base should be blue-green or green. The color of end point may be judged by the potentiometric titration method, and a blank titration is repeated to decrease the titration error.

Methyl violet, α-naphthalphenol benzyl alcohol and quinidine red are other indicators for titrating weak bases in HAc solvent. The acidic color of 0.5% methyl violet in HAc solvent is blue, and its basic color is purple. The acidic color of 0.2% α-naphthalphenol benzyl alcohol in HAc solvent is green, and its basic color is yellow, while the acidic color of 0.1% quinidine red in methanol solvent is colorless, and its basic color is red.

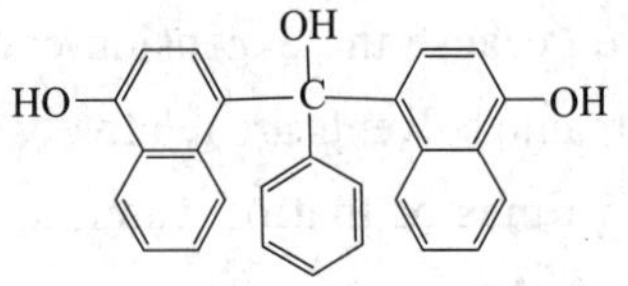

α-Naphthalphenol benzyl alcohol

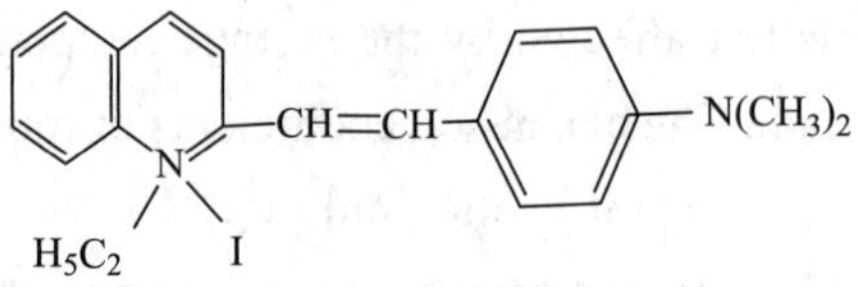

Quinaldine red

In the non-aqueous titration, the end points of titrating many substances may be determined by the potentiometry when there is no suitable indicator.

2.1.4 Application examples

Most compounds containing basic groups can be titrated with standard $HClO_4$ solution, such as amines, amino acids, heterocyclic compounds containing nitrogen, organic base, etc.

(1) Organic weak base. Organic weak base, which dissociation constants in water (K_b) is over 10^{-11}, is usually titrated with $HClO_4$-HAc standard solution.

For example, determination of the content of alkaloid: weigh accurately 40−100mg sample, add 5ml anhydrous acetic acid to dissolve the sample, add 1 drop of crystal violet, titrate the sample solution with $HClO_4$-HAc standard solution to the green end point. The content of alkaloid in the sample is calculated with the volume of $HClO_4$-HAc standard solution added.

$$\text{Alkaloid\%} = \frac{c_{HClO_4} \cdot V_{HClO_4} \cdot M_{alkaloid}}{n \times m_S \times 1000} \times 100\%$$

(2) Hydrohalites of organic bases. Organic bases, poor soluble in water and unstable, usually become their salts after reacting with acids for medicinal function, and most of them are organic alkali

hydrohalites (B · HX), such as ephedrine hydrochloride, ligustrazine hydrochloride, berberine hydrochloride and Scopolamine hydrobromide, etc. The above medicines can all be titrated in non-aqueous solution. Organic alkali hydrohalites cannot be titrated directly with $HClO_4$ because of the stronger acidity of halogen acid in acetic acid. Usually excessive $Hg(Ac)_2$-HAc solution is added, halogen acid is converted into HgX_2 which is difficult to ionize, and hydrohalites are converted into weaker acetates, and the weaker acetates can be titrated with $HClO_4$-HAc standard solution.

$$2B \cdot HX + Hg(Ac)_2 \rightleftharpoons 2B \cdot HAc + HgX_2$$

$$B \cdot HAc + HClO_4 \rightleftharpoons B \cdot HClO_4 + HAc$$

For example, ephedrine hydrochloride is the conjugated acid of ephedra alcaloids.

Procedure for determining the content of ephedrine hydrochloride: weigh accurately 0.1-0.15g sample of ephedrine hydrochloride, add 10ml HAc and dissolve the sample by heating, add 4ml $Hg(Ac)_2$ reagent and 1 drop indicator of crystal violet, titrate the solution with 0.1mol/L $HClO_4$-HAc standard solution until the blue-green end point, and repeat the blank titration to calibrate the volume of the titrant consumed, calculate the content of ephedrine hydrochloride according to the volume of the titrant consumed.

2.2 Titration of acids

2.2.1 Non-aqueous solvents

The acid substances insoluble or with the dissociation constant below 10^{-7} in aqueous solution, cannot be titrated directly with standard NaOH solution. The strength of a weak acid increases if it is placed in a solvent that is more basic than water, and it can be titrated with standard basic solution. Usually the carboxylic acids are titrated in alcohol solvents, while the weaker or extremely weak acid in alkaline solvents that are more basic than water, such as ethylenediamine and dimethylformamide, etc. Differentiating solvents, eg. methyl isobutyl ketone, are usually selected as the solvent medium for titrating the mixed acid separately. Some mixed solvents or aprotic solvents are often used, too.

2.2.2 Standard solution and primary standard substances

The standard basic solution to titrate acids are alcoholic-alkalines (sodium methoxide, potassium methoxide, lithium methoxide, sodium amino ethoxide, and so on), alkaline hydroxides (potassium hydroxides, sodium acetate, KHP, dimethyl sulfoxide and sodium triphenylmethane, etc), Quaternary ammonium hydroxides (tetrabutylammonium hydroxides, tributylammonium hydroxides, triethylbutylammonium hydroxides, etc). The widely used titrant is sodium methoxide in benzene-methanol mixed solvents.

(1) Preparation of 0.1mol/L sodium methoxide in benzene-methanol standard solution: 150ml anhydrous methanol containing H_2O% below 0.2% is taken and placed the container cooled in ice-water, 2.5g metallic sodium is repeatedly added in small quantities, anhydrous benzene containing H_2O% below 0.2% is added into the solution up to 1000ml after metallic sodium completely dissociated.

Note that methanol is the conjugated acid of sodium methoxide, too much methanol added in this standard solution is not suitable, or the alkality of the standard solution decreases. Metallic sodium reacts with anhydrous methanol violently, so adequate cooling is necessary during the preparation of this standard solution to ensure safety.

$$2CH_3OH + 2Na \rightleftharpoons 2CH_3ONa + H_2\uparrow$$

(2) Calibration of 0.1mol/L sodium methoxide in benzene-methanol standard solution: benzoic acid

is the commonly used primary standard substance to calibrate 0.1mol/L sodium methoxide in benzene-methanol standard solution. Thymol blue as the indicator, benzoic acid is titrated with sodium methoxide to the blue end point. Repeat the blank titration, V_0 is the volume of standard CH_3ONa solution consumed in the bland titration.

$$C_6H_5COOH + CH_3ONa \rightleftharpoons C_6H_5COO^- + CH_3OH + Na^+$$

$$c_{CH_3ONa} = \frac{m_{C_6H_5COOH}}{(V_{CH_3ONa} - V_0) \cdot M_{C_6H_5COOH}} \times 1000$$

2.2.3 Indicators

(1) Thymol blue. Thymol blue is a suitable indicator for titrating medium-strong acids in the solvent of benzene or butylamine, pyridine, dimethyl formamide, tertiary butanol, etc. Thymol blue can give a sharp end point, and its acidic color is yellow, and its basic color is blue.

(2) Azo violet. Azo violet is used for titrating weaker acids in the solvent of butylamine or ethylenediamine, dimethyl formamide, pyridine, acetonitrile, ketones and alcohols, etc. The acidic color of azo violet is red, and its basic color is blue.

(3) Bromophenol blue. Bromophenol blue is usually used to titrate carboxylic acids in the solvent of benzene or methanol, trichlormethane, etc. The acidic color of Bromophenol blue is yellow, and its basic color is blue.

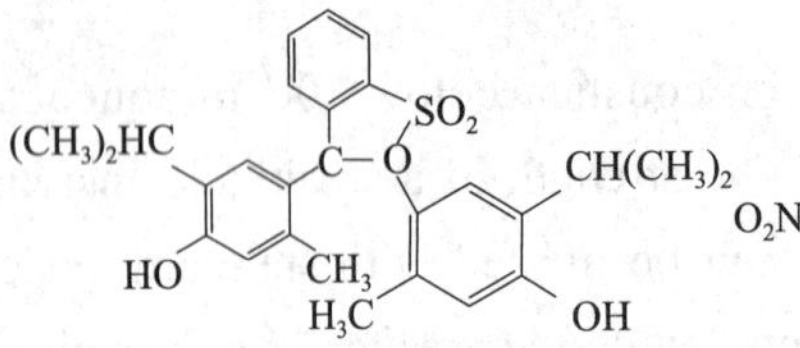

Thymol blue 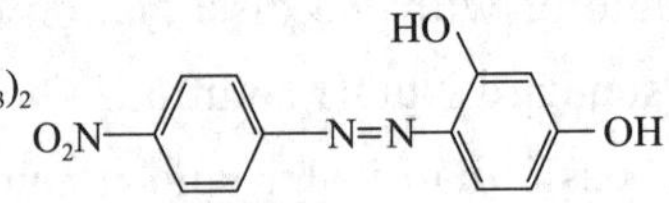　　Azo violet 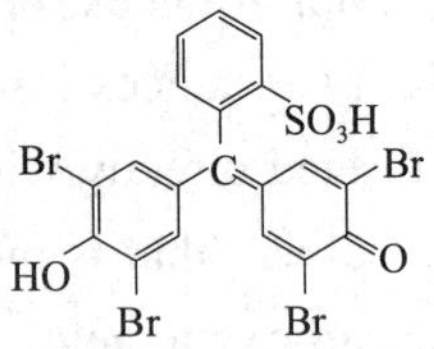　　Bromophenol blue

2.2.4 Application examples

(1) Carboxylic acids. The carboxylic acids, which dissociation constants are between 10^{-5}-10^{-6} in aqueous solutions, can be titrated with standard NaOH solution in alcohol solvent, and phenolphthalein as the indicator; Weaker carboxylic acids with smaller dissociation constants in aqueous solutions are usually titrated with sodium methoxide standard solution in benzene-methanol solvent, and thymol blue as the indicator. The reaction between the weak carboxylic acids and sodium methoxide is as follow.

$$RCOOH + CH_3ONa \rightleftharpoons RCOONa + CH_3OH$$

Procedure: add 20–30ml benzene-methanol (4 : 1) into a conical flask, add 2–3 drops 0.3% thymol blue, titrate the above solution with standard sodium methoxide solution to the blue end point, record the volume of standard sodium methoxide solution consumed. Then weigh accurately 0.5–0.8g sample of carboxylic acid into the same conical flask, repeat the above procedures in the blank titration, and record the volume of standard sodium methoxide solution consumed. The content of carboxylic acid is calculated according to the volume of standard sodium methoxide solution consumed.

$$\text{Carboxylic acid\%} = \frac{c_{CH_3ONa} \cdot (V - V_0) \cdot M_{RCOOH}}{n \times m_S \times 1000} \times 100\%$$

n is the moles of sodium methoxide reacting with 1mol carboxylic acid, V and V_0 are respectively the volumes of standard sodium methoxide solution assumed to titrate the sample and the blank solution (ml).

(2) Phenolic compounds. Phenolic compounds are weaker than the corresponding carboxylic acids, for example the dissociation constant of benzoic acid in aqueous solution is $10^{-4.2}$, while the dissociation constant of phenol is only $10^{-9.96}$ and cannot be titrated in aqueous solution because of lacking a distinct end point. The acidity of phenol will increase if it is placed in an ethylenediamine solvent, and it can be titrated with a standard $NH_2CH_2CH_2ONa$ solution, in which there is an obvious end point. The acidity of phenolic compounds will increase if they ortho-substituted or para-substituted by nitro-, aldehyde group, bromo-, chloro-, etc. The above phenolic compounds ortho-substituted or para-substituted can be titrated with standard sodium methoxide solution in dimethyl formamide, and azo violet as the indicator. For example, the content of gossypol in the cottonseed or cottonseed oil is determined according to the above titration procedures in aqueous solution.

重点小结

在酸碱质子论中，能给出质子的物质称为酸，能接受质子的物质称为碱。水溶液中，酸碱存在得失质子平衡，酸碱与其得失质子的产物组成共轭酸碱对，一对共轭酸碱对仅相差一个质子。酸碱反应的实质是质子在两对共轭酸碱对间转移。

溶液中酸碱平衡后，酸碱各型体的浓度和分布系数均由溶液的pH决定，溶液中氢离子浓度的计算，在5%的误差下可以采用最简式来近似计算，从而得到酸碱滴定过程中溶液pH随着滴定剂的体积的变化规律，即酸碱滴定曲线。在酸碱滴定中，化学计量点pH由滴定反应所生成的产物决定。化学计量点前后0.1%，pH发生了急剧变化称为滴定突跃，酸碱滴定突跃的大小与酸碱的强度和溶液的浓度有关，酸（碱）越强，滴定突跃越大；溶液浓度越大，滴定突跃越大。只有酸（碱）的$c \cdot K_a \geqslant 10^{-8}$或$c \cdot K_b \geqslant 10^{-8}$，才能进行准确滴定。对于多元酸（碱）$c \cdot K_{ai} \geqslant 10^{-8}$（$c \cdot K_{bi} \geqslant 10^{-8}$），此步酸（碱）电离可以准确滴定，$\dfrac{K_{a_i}}{K_{a_{i+1}}} \geqslant 10^4 \left(\dfrac{K_{b_i}}{K_{b_{i+1}}} \geqslant 10^4\right)$，此步酸（碱）解离与下一级解离是分开的，可以完成此级解离的分步滴定。

酸碱指示剂是有机弱酸或弱碱，在溶液中不完全电离，存在酸式型体活碱式型体，而且两种型体存在不同的颜色。随着溶液pH的改变酸式型体变成碱式型体，或由碱式型体变成酸式型体，呈现不同的颜色。$pH<pK_{HIn}-1$，指示剂显酸式色，$pH>pK_{HIn}+1$，指示剂显碱式色，$pK_{HIn}-1 \leqslant pH \leqslant pK_{HIn}+1$，是指示剂的变色范围。指示剂的变色范围受温度、指示剂用量、溶剂、溶液离子强度、滴定顺序等因素影响。酸碱滴定突跃是指示剂选择的依据。指示剂变色范围全部或部分落在滴定突跃范围内。

酸碱滴定中化学计量点与滴定终点不一致所造成的终点误差（TE%），可以用滴定终点的pH来估算。强酸滴定强碱的终点误差$TE\% = \dfrac{[H^+]_{ep}-[OH^-]_{ep}}{c_{ep}} \times 100\%$，强碱滴定强酸的终点误差$TE\% = \dfrac{[OH^-]_{ep}-[H^+]_{ep}}{c_{ep}} \times 100\%$，强酸滴定弱碱的终点误差$TE\% = \dfrac{[H^+]_{ep}-([OH^-]_{ep}+[A^-]_{ep})}{c_{ep}} \times 100\%$，强碱滴定弱酸的终点误差$TE\% = \dfrac{[OH^-]_{ep}-([H^+]_{ep}+[HA^-]_{ep})}{c_{ep}} \times 100\%$。

最常用的酸性标准溶液和碱性标准溶液分别是0.1mol/L HCl和0.1mol/L NaOH，都采用标定法配制，可以对含有酸性或碱性基团的多种物质进行滴定分析。

目标检测

1. (1) Write down the conjugate acids of the following bases:

HCO_3^-, CH_3COO^-, H_2O, $C_6H_5NH_2$, NH_3, Ac^-, S^{2-}

(2) Write down the conjugate bases of the following acids:

HNO_3, H_2O, $H_2PO_4^-$, HCO_3^-, $HC_2O_4^-$, H_2S, HPO_4^{2-}

2. Explain the acid-base dissociation, salt hydrolysis and the essence of acid-base neutralization reaction with the acid-alkali proton theory.

3. Write the proton balance equation for the following substances in aqueous solution.

HNO_3 HCN NH_3 NH_4 NH_4Ac Na_2HPO_4 Na_3PO_4 H_2CO_3 H_3PO_4

4. Calculate the pH of the following solutions.

(1) 0.10mol/L NaAc ($K_{a(HAc)}$=1.8×10^{-5})

(2) 0.10mol/L NH_4Cl ($K_{b(NH_3)}$=1.8×10^{-5})

5. Describe the discoloration principle of acid-base indicators, the ranges of color changes and the principle of selecting indicators?

6. How to explain the difference between the actual transition ranges and the theoretical transition ranges of methyl orange?

7. Explain the influence of water on acid-base titration in nonaqueous solution with the acid-alkali proton theory.

8. Why are the standard solutions used in neutralization titrations generally strong acids and bases rather than weak acids and bases?

9. What is acid–base titration curve and pH jump range? What factors affect end-point sharpness in an acid or base titration?

10. Consider curves for the titration of 0.10mol/L HCl and 0.010mol/L HAc with 0.10mol/L NaOH.

(1) Briefly account for the differences between curves for the two titrations.

(2) In what respect will the two curves be indistinguishable?

11. Whether can the following acid (base) be titrated with 0.1000mol/L NaOH (HCl)? If can, which indicators can be selected?

(1) 0.1000mol/L HCOOH (K_a=1.8×10^{-4})

(2) 0.1000mol/L NH_4Cl (K_b =1.8×10^{-5})

(3) 0.1000mol/L C_6H_5COOH (K_a= 6.3×10^{-5})

(4) 0.1000mol/L C_6H_5COONa

(5) 0.1000mol/L C_6H_5OH (K_a=1.0×10^{-10})

(6) 0.1000mol/L C_6H_5ONa

12. Whether can the following acid (base) be titrated with 0.1000mol/L NaOH (HCl) separately?

(1) 0.10mol/L H_3PO_4

(2) 0.10mol/L $H_2C_2O_4$

(3) 0.10mol/L H_2SO_4 + 0.10mol/L H_3BO_3

(4) 0.10mol/L NaOH + 0.10mol/L $NaHCO_3$

13. Why is it common practice to boil the solution near the equivalence point in the standardization of Na_2CO_3 with acid?

14. In which of the following solvents, the acidity strength of acetic acid, benzoic acid, hydrochloric acid, perchloric acid are the same?

(1) pure water (2) concentrated sulfuric acid (3) liquid ammonia
(4) methyl isopropyl ketone (5) ethanol

15. Which is protonic solvent in the following solvents? Which is the aprotic solvent? If it is protonic solvent, acid solvent or alkaline solvent? If it is aprotic solvent, inert solvent or alkaline aprotic solvent?

(1) methyl isobutyl ketone (2) benzene (3) water
(4) acetic acid (5) ethylenediamine (6) 1,4-dioxane
(7) diethyl ether (8) isopropanol (9) butylamine
(10) acetone

16. Does the NaOH standard solution absorbed CO_2 in the air have any influence on the determination results when it is used for the determination of strong acid and weak acid?

17. Why can hydrochloric acid be used to titrate borax instead of sodium acetate? Why can sodium hydroxide titrate acetic acid instead of boric acid?

18. Calculate the pH of the following solutions.

(1) 0.10mol/L NaH_2PO_4

(2) 0.05mol/L HAc + 0.05mol/L NaAc

19. Calculate the pH of a solution prepared by mixing 2.0ml of a strong acid of pH 3.00 and 3.0ml of a strong base of pH 10.00.

20. The autoprolysis constant of water and ethanol: $K_s^{H_2O}=1.0\times10^{-14}$, $K_s^{C_2H_2OH}=1.0\times10^{-19.1}$.

(1) What is the pH of pure water and the $pC_2H_5OH_2$ of ethanol?

(2) What is the pH, $pC_2H_5OH_2$ and pOH, pC_2H_5O of aqueous solution and ethanol solution of 0.0100mol/L of $HClO_4$? (It is assumed that $HClO_4$ is completely dissociated.)

21. Take 1.250g of pure HA to prepare 50ml aqueous solution, 41.10ml of NaOH solution (0.0900mol/L) was consumed and titrated it to the stoichiometric point. During the titration, when the titrant was added to 8.24ml, the pH of the solution was 4.30. Calculate:

(1) molar mass of HA

(2) the K_a value of HA

(3) pH of stoichiometric point

22. A 0.3000g feed sample is analyzed for its protein content by the modified Kjeldahl method. If 25.0ml of 0.100mol/L HCl is required for titration, what is the percent protein content of the sample? (M_N=14.01)

23. A 0.2027g sample of finely powered limestone (mainly $CaCO_3$) was dissolved in 50.00ml of 0.1035mol/L HCl. The solution was heated to expel CO_2 produced by the reaction. The remaining HCl was then titrated with 0.1018mol/L of NaOH and it required 16.62ml. Calculate the percentage of $CaCO_3$ in the limestone sample. (M_{CaCO_3}=100.09g/mol)

24. A mixture of HCl and H_3PO_4 is titrated with 0.1000mol/L NaOH. The first end point (methyl red) occurs at 35.00ml, and the second end point (bromthymol blue) occurs at a total of 50.00ml (15.00ml after the first end point). Calculate the millimoles HCl and H_3PO_4 present in the solution.

25. A 0.9872g sample of unknown potassium hydrogen phthalate (KHP) required 28.23ml of 0.1037mol/L NaOH for neutralization. What is the percentage of KHP in the sample? (M_{KHP}=202.4g/mol)

26. A sample that may be a mixture of Na_2CO_3 and $NaHCO_3$ or NaOH and Na_2CO_3 is titrated with the double-indicator method. A 0.2075g sample required 35.84ml of 0.1037mol/L HCl to reach the phenolphthalein end point, and an additional 5.96ml to reach the bromocresol green end point. Determine the mixtures and calculate the percentage of each component. (M_{NaOH}=39.997g/mol, $M_{Na_2CO_3}$=105.99g/mol)

27. 0.1000mol/L NaOH is titrated with 0.1000mol/L HCl,

(1) Methyl orange as the indicator, the pH at the end point is 4.0,

(2) The mixed indicator of methyl red- bromocresol green is used, the pH at the end point is 5.0, calculate their titration error respectively.

28. 0.1000mol/L HAc is titrated with 0.1000mol/L NaOH,

(1) Neutral red as the indicator, the pH at the end point is 7.0,

(2) Phenolphthalein as the indicator, the pH at the end point is 10.0, calculate their titration error respectively.

29. If 1.000L of 0.1500mol/L NaOH was unprotected from the air after standardization and absorbed 11.2mmol of CO_2, what is its new molar concentration when it is standardized against a standard solution of HCl using

(1) phenolphthalein?

(2) bromocresol green?

30. The molar concentration of $HClO_4$-HAc solution is 0.1086mol/L, when calibrated at 24℃, calculate its molar concentration at 30℃.

31. 0.2500g alkaloid sample containing amino group is dissolved in acetic acid solvent, and is titrated with 0.1000mol/L $HClO_4$-HAc solution, and 12.00ml titrant is consumed, calculate the content of amino group in the sample. (M_{NH_2}=16.033g/mol)

32. Weigh accurately 0.1498g ephedrine hydrochloride, dissolve the sample in 10ml HAc, add 4ml $HgAc_2$ and 1 drop crystal violet, 8.02ml 0.1003mol/L $HClO_4$ solution consumed, 0.65ml standard $HClO_4$ solution consumed in the blank titration, calculate the mass percent of ephedrine hydrochloride in the sample. ($M_{C_{10}H_{15}NO \cdot HCl}$=201.7g/mol)

Chapter 6　Precipitation Titration

学习目标

知识要求

1. **掌握**　银量法三种指示剂的基本原理、滴定条件和标准溶液的配制与标定。
2. **熟悉**　沉淀滴定法滴定曲线的绘制。
3. **了解**　沉淀滴定法在药物分析中的应用。

能力要求

通过学习沉淀滴定法基本原理，银量法的滴定条件以及在药物分析中的应用，初步掌握沉淀滴定法方法运用，能够根据样品性质，设计适合的沉淀滴定方法及基本操作方案。

Section 1　Introduction to Precipitation Titration

PPT

Precipitation titration (沉淀滴定法) is one of the method of titrimetric analysis based on the precipitation reactions. There are many reactions which produce insoluble precipitates, the requirements of precipitation titration should be as follows: ① the solubility (s) of precipitate is small enough to assure the completeness of the precipitation reaction; ② the reaction should be stoichiometric, and the stoichiometry must be exact; ③ the rate of titrimetric equilibria must be sufficiently rapid; ④ an available mean of detecting the stoichiometric point is available.

Section 2　Argentometry

PPT

1. Basic principles

The precipitation reaction based on the formation of insoluble silver salts are commonly used in titrimetry. Such as:

$$Ag^+ + X^- \rightleftharpoons AgX\downarrow \qquad (X^-: Cl^-, Br^-, I^-, SCN^-, etc.)$$

The titrimetric analysis based on this type of precipitation reaction is termed argentometry (银量法). Argentometry determines the compounds with chloride (Cl^-), bromide (Br^-), iodide (I^-), thiocyanate (SCN^-) and silver (Ag^+) by a direct or indirect procedure. And the organic compounds also can be determined by the treatment of transformation.

Argentometry is most widely used while several others precipitation methods are developed, such as the reaction between sodium tetraphenylboron ($NaB(C_6H_5)_4$) and K^+, potassium ferrocyanide ($K_4[Fe(CN)_6]$) and Zn^{2+}, Ba^{2+}, Pb^{2+} and SO_4^{2-}. This paper mainly describes the basic principle and the application of argentometry.

1.1 Titration curves

Consider the titration of 20.00ml of 0.1000mol/L NaCl with 0.1000mol/L $AgNO_3$, a titration curve, constructed by plotting the ion concentration of Cl^- (pCl) against percent titration ($T\%$) or volume (V) of $AgNO_3$ added, gives a useful indication of the sharpness of a titration reaction.

(1) Initial Point

The concentration of Cl^- ($[Cl^-]$) is determined by NaCl all in colution:

$$[Cl^-] = 0.1000\text{mol/L} \qquad pCl=-\lg 0.1000 = 1.00$$

(2) Before the stoichiometric point

$[Cl^-]$ is determined by remaining NaCl in solution. When V ml of $AgNO_3$ is added to precipitate the solution, the $[Cl^-]$ in solution can be calculated as follows:

$$[Cl^-] = \frac{(20.00-V)\times 0.1000}{(20.00+V)}$$

For example, when 19.98ml of $AgNO_3$ is added:

$$[Cl^-] = \frac{(20.00-19.98)\times 0.1000}{(20.00+19.98)} = 5.0\times 10^{-5}\text{mol/L} \quad pCl=4.30$$

$$K_{sp} = [Ag^+][Cl^-] = 1.77\times 10^{-10} \quad pAg + pCl = -\lg K_{sp} = 9.74$$

$$pAg = 9.74-4.30 = 5.44$$

(3) At the stoichiometric point

$[Cl^-]$ is determined by saturated solution of AgCl:

$$[Ag^+] = [Cl^-] = \sqrt{K_{sp}} = \sqrt{1.8\times 10^{-10}} = 1.34\times 10^{-5}\text{mol/L}$$

$$pAg = pCl = \frac{1}{2}pK_{sp} = 4.87$$

(4) Beyond the stoichiometric point

$[Cl^-]$ is determined by excess $AgNO_3$ added:

$$[Ag^+] = \frac{(20.02-20.00)\times 0.1000}{(20.00+20.02)} = 5.0\times 10^{-5}\text{mol/L}$$

$$pAg = 4.30 \quad pCl = 9.74 - 4.30 = 5.44$$

The titration curve of NaBr and I can be calculated and plotted in the same way. Additional data with pX or pAg as the y-axis and $T\%$ or V as the x-axis are provided in Table 6-1. The titration curve according to these data is given in Figure 6-1 and Figure 6-2.

Table 6-1 Changes during the titration of 20.00ml of NaX with $AgNO_3$

$AgNO_3$ (0.1000mol/L)		Cl^- (0.1000mol/L)		Br^- (0.1000mol/L)		I^- (0.1000mol/L)	
V (ml)	percentage of titration (%)	pCl	pAg	pBr	pAg	pI	pAg
0.00	0.0	1.00	—	1.00	—	1.00	—
18.00	90.0	2.28	7.46	2.28	10.02	2.28	13.75
19.00	98.0	3.00	6.74	3.00	9.30	3.00	13.03
19.80	99.0	3.30	6.44	3.30	9.00	3.30	12.73
19.96	99.8	4.00	5.74	4.00	8.30	4.00	12.03
19.98	99.9	4.30	5.44	4.30	8.00	4.30	11.73
20.00	100.0	4.87	4.87	6.15	6.15	8.02	8.02
20.02	100.1	5.44	4.30	8.00	4.30	11.73	4.30
20.04	100.2	7.74	4.00	8.30	4.00	12.03	4.00
20.20	101.0	6.44	3.30	9.00	3.30	12.73	3.30
20.40	102.0	6.74	3.00	9.30	3.00	13.03	3.00
22.00	110.0	7.42	2.32	9.98	2.32	13.71	2.32
40.00	200.0	8.27	1.48	10.82	1.48	14.55	1.48

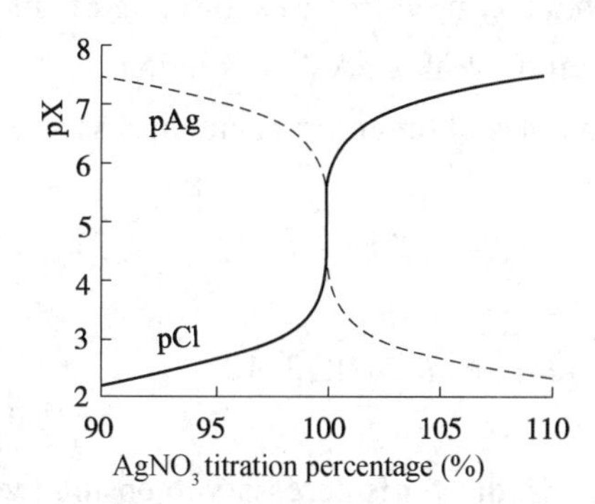

Figure 6-1 Titration curve of NaCl with $AgNO_3$ (0.1000mol/L)

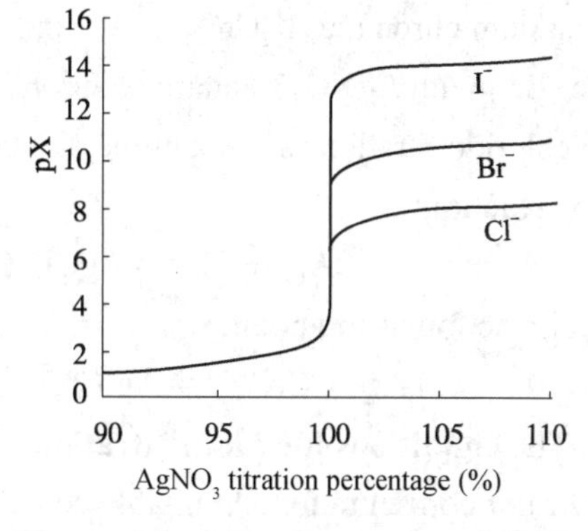

Figure 6-2 Titration curve of NaCl, NaBr and NaI with $AgNO_3$ (0.1000mol/L)

The titration curve of precipitation titration is closely analogous to that of a strong acid-strong base titration:

(1) The titration curve of Cl^- and the titration curve of Ag^+ is roughly symmetrical to the stoichiometric point. $[Cl^-]$ decreased in same rate while $[Ag^+]$ increased, two curves intersect at the stoichiometric point: $[Ag^+] = [Cl^-]$.

(2) Similar to the titration curve of acid-base titration: at the beginning, the remaining $[X^-]$ is still high enough, although part of the X^- are removed from solution by precipitation as AgX, pX increases slowly. Near the stoichiometric point, the remaining $[X^-]$ is smaller and smaller, pX has a remarkable change with a very small volume of Ag^+ added, it forms rough sudden titration break.

(3) The titration break depend on both the concentration of solution (*c*) and solubility product constants (K_{sp}). ① The smaller the K_{sp} is, the longer the break is. For example, to approximate equimolar

mixtures of AgI, AgBr and AgCl, the break shows in order: ΔpI > ΔpBr > ΔpCl corresponding to the K_{sp}, $K_{sp\,(AgI)} < K_{sp\,(AgBr)} < K_{sp\,(AgCl)}$. ② The smaller the c is, the shorter the break is.

1.2 Step by step titration of halogen ions

Step by step titration is convenient to the titration of an approximate equimolar mixture of three halides (Cl^-, Br^-, I^-) with $AgNO_3$ standard solution because of widely differing of three precipitates. AgI precipitates first and AgCl precipitates last corresponding to the K_{sp} of AgX, $K_{sp\,(AgI)} < K_{sp\,(AgBr)} < K_{sp\,(AgCl)}$. Titration curve of mixtures shows three stoichiometric points and three titration breaks with qualitatively similar distortions. Actually, fractional precipitation is usually not very efficient. Due principally to adsorption on the surface of the AgX precipitate and to mixed crystal formation in titration. This method does not provide as good a separation as theory would indicate for the increased error.

2. Methods of end point detection

Three types of indicator methods are commonly applied to the argentometry titration of Cl^- with Ag^+: ① Mohr method (莫尔法) with potassium chromate (K_2CrO_4) indicator. ② Volhard method (佛尔哈德法) with ammonium ferric sulfate ($NH_4Fe(SO_4)_2 \cdot 12H_2O$) indicator. ③ Fajans method (法扬司法) with adsorption indicator.

2.1 Mohr method

2.1.1 Reaction principle

The Mohr method, based on the formation of orange-red silver chromate (Ag_2CrO_4) at the end point by using potassium chromate (K_2CrO_4) as the indicator, is mainly used for direct titrating of chloride (Cl^-) and bromide (Br^-) with $AgNO_3$ standard solution in a neutral or weakly alkaline solution.

Taking chloride titration as an example, the basic principles of the reaction are discussed below:

Titration reaction:

$$Ag^+ + Cl^- \rightleftharpoons AgCl\downarrow \text{ (white)} \qquad K_{sp} = 1.77\times10^{-10}$$

Indicator reaction at end point:

$$2Ag^+ + CrO_4^{2-} \rightleftharpoons Ag_2CrO_4\downarrow \text{ (orange-red)} \qquad K_{sp} = 1.12\times10^{-12}$$

2.1.2 The conditions for Mohr titration

(1) Indicator concentration: A suitable concentration of indicator is necessary to ensure the accuracy of the results. Negative error occurs because Ag_2CrO_4 precipitates earlier than expected when there are too much indicator, and positive error occurs because the color of Ag_2CrO_4 is too light to observe when there are too little indicator. The yellow color of the indicator also interferes with the observation of orange-red color at end point.

It is appropriate to add 1–2ml of 5% (g/ml) indicator into 50–100ml of solution, the suitable indicator concentration is 2.6×10^{-3}–5.2×10^{-3}mol/L.

(2) Acidity: A serious limitation of the Mohr method is that the solution should be neutral or weak alkalinity, pH is 6.5–10.5, to ensure the formation of Ag_2CrO_4. Below pH 6.5, the solubility of Ag_2CrO_4 becomes excessive, that means the process of precipitation is delayed or absent.

$$Ag_2CrO_4 + H^+ \rightleftharpoons 2Ag^+ + HCrO_4^-$$

$$2CrO_4^{2-} + 2H^+ \rightleftharpoons 2HCrO_4^- \rightleftharpoons Cr_2O_7^{2-} + H_2O$$

Coprecipitation of silver hydroxide (Ag_2O) occurs above pH 10.5.

$$2Ag^+ + 2OH^- \rightleftharpoons 2AgOH \rightleftharpoons Ag_2O\downarrow \text{ (black)} + H_2O$$

The presence of ammonium salts may cause a significant error because of the formation of $[Ag(NH_3)_2]^+$ affecting the solubility of AgCl and Ag_2CrO_4. So the pH should be 6.5–7.2 to prevent the formation of $[Ag(NH_3)_2]^+$ when there are ammonium salt in solution.

Sodium carbonate (Na_2CO_3), calcium carbonate ($CaCO_3$) or disodium tetraborate ($Na_2B_4O_7 \cdot 10H_2O$) can be used conveniently to neutralize excess acid when acidity is too high, and dilute nitric acid (HNO_3) can be added conveniently when acidity is too low.

(3) Vigorous shaking: Vigorous shaking is rather effective for releasing Ag^+ adsorbed by AgCl precipitation strongly.

(4) Pre-separation: It is a useful method to remove the interfering ions from solution before titrating, such as ions formed chromate precipitates (Ba^{2+}, Pb^{2+}, Hg^+) or silver precipitation (SO_3^{2-}, PO_4^{3-}, AsO_4^{3-}, S^{2-}, $C_2O_4^{2-}$), colored ions (Cu^{2+}, Co^{2+}, Ni^{2+}) and ions were prone to hydrolysis (Fe^{3+}, Al^{3+}), etc.

2.1.3 Application

The Mohr method is especially widely used for bromide (Br^-) and chloride (Cl^-) by direct titration, and it also used for cyanide (CN^-) in weakly alkaline solution. This method is unfit for iodide (I^-) and thiocyanate (SCN^-) because excessive adsorption of AgI and AgSCN to make the end point occur earlier than expected.

2.2 Volhard method

2.2.1 Reaction principle

The Mohr method, based on the formation of red $[Fe(SCN)]^{2+}$ at the end point by using ammonium ferric sulfate ($NH_4Fe(SO_4)_2 \cdot 12H_2O$) as the indicator, is widely used for direct titrate of Ag^+ with NH_4SCN or KSCN standard solution and also for back-titrate of X^- with NH_4SCN and $AgNO_3$ standard solution in a strong acidic solution.

(1) Direct titration

Ag^+ is titrated with SCN^- in the presence of Fe^{3+} indicator first, slightly excess SCN^- form red colored $[Fe(SCN)]^{2+}$ with Fe^{3+} indicator at the end point of the titration.

Titration reaction: $Ag^+ + SCN^- \rightleftharpoons AgSCN\downarrow$ (white)

Indicator reaction at end point: $Fe^{3+} + SCN^- \rightleftharpoons [Fe(SCN)]^{2+}\downarrow$ (red)

(2) Back-titration

The procedure is the addition of an excess of $AgNO_3$ to a solution of halide first, then the addition of $NH_4Fe(SO_4)_2 \cdot 12H_2O$ indicator after the formation of AgX, followed by the back-titration of the remaining $AgNO_3$ with NH_4SCN.

Titration reaction: Ag^+ (excess) $+ X^- \rightleftharpoons AgX\downarrow$

Ag^+ (surplus) $+ SCN^- \rightleftharpoons AgSCN\downarrow$ (white)

Indicator reaction at end point: $Fe^{3+} + SCN^- \rightleftharpoons [Fe(SCN)]^{2+}\downarrow$ (red)

NH_4SCN can slowly convert AgCl to AgSCN precipitate close to the end point as the reaction below:

$$AgCl \rightleftharpoons AgSCN\downarrow \text{(red)} + Cl^-$$

Therefore, removing AgCl before titrating is necessary to reduce the positive titration error caused to the more consume of SCN^- than expected: ① Pre-separation by filtration: To avoid the convert reaction, AgCl precipitate must be separated before titrating. First, Most of the adsorbed Ag^+ is removed from AgCl surface by boiling the suspension for a few minutes to coagulate the silver chloride before filtration. Second, Ag^+ remained in the AgCl surface is removed by filtering and washing AgCl with diluted HNO_3. Then, the excess Ag^+ in collected wash solution is titrated with NH_4SCN standard solution. This

method may cause some error due to a cumbersome precipitation and filtration scheme. ② Without pre-separation: 1–3ml of nitrobenzene, an immiscible organic solvent, is added to the suspension to coat the AgCl particles, thereby protecting them from interaction with KSCN and preventing them from being converted to AgSCN.

NH_4SCN does not convert AgBr or AgI to AgSCN precipitate because the solubility of AgBr and AgI is smaller than AgSCN.

2.2.2 The conditions for Volhard titration

(1) The titration must be carried out in a strongly acidic solution to prevent hydrolysis of Fe^{3+}.

(2) Vigorous shaking is effective for releasing Ag^+ adsorbed strongly by AgSCN precipitation in direct titration.

(3) Filtration of AgI or addition of nitrobenzene are useful to prevent conversion of AgCl in back-titration.

(4) The operation should be performed at room temperature. End point cannot be indicated because of the discoloration of the red $[Fe(SCN)]^{2+}$ at high temperature.

(5) Excess AgCl standard solution must be added before $NH_4Fe(SO_4)_2 \cdot 12H_2O$ indicator to prevent oxidation of I^- with Fe^{3+} in back-titration.

2.2.3 Application

The Volhard method is widely used for Ag^+ by direct titration and for Cl^-, Br^-, I^-, SCN^-, PO_4^{3-}, AsO_4^{3-} by indirect titration or back-titration due to good selectivity and less interference because many interference ions, such as PO_4^{3-}, CO_3^{2-}, AsO_4^{3-}, rarely react with Ag^+ in a strongly acidic solution.

2.3 Fajans method

2.3.1 Reaction principle

Fajans method is used for titrating of X^- with $AgNO_3$ standard solution, and the special adsorption indicator is used for end point indication.

Adsorption indicator, a kind of organic dye, exhibits a certain color when it is dissociated into ions in solution. A change in color, attributed to the strong deformation of indicator ions by the adsorption of dye ions on colloidal precipitate surface with opposite charge, should occur just at the isoelectric point thereby indicating the end point. There are two kinds of adsorbed indicator: anionic dye and cationic dye. An anionic dye, a kind of organic weak acid such as fluorescein and its derivatives, shows anionic structure in solution and a cationic dye, a kind of organic weak base such as methyl violet and rhodamine 6G, shows cationic structure in solution.

Consider the use of fluorescein (HFI) as an indicator for the titration of chloride (Cl^-) with $AgNO_3$ standard solution, the basic principles of the reaction are illustrated below:

Fluorescein indicator is dissociated into an anionic ion (FI^-) in solution. Before the end point, AgCl colloidal precipitate is negatively charged on its surface owing to the adsorption of excess Cl^- and this anionic formation can be marked as $(AgCl) \cdot Cl^-$. FI^- is repelled by $(AgCl) \cdot Cl^-$ because of the similarly negatively charger and is remained in solution, thereby the color of solution is yellow-green. After the end point, when the isoelectric point is passed, AgCl colloidal precipitate turns to positively charged owing to the adsorption of excess Ag^+ and this cationic formation can be marked as $(AgCl) \cdot Ag^+$. FI^- is immediately adsorbed on the surface of $(AgCl) \cdot Ag^+$ to form an adsorbed silver salt of the dye marked as $(AgCl) \cdot Ag^+ \cdot FI^-$, and the change of the color from yellow-green to pink, attributed to the strong deformation of FI^- by the adsorbed Ag^+, just indicates the end point.

This process of color changing is illustrated as below:

$$HFI \rightleftharpoons H^+ + FI^- \quad \text{(yellow-green)}$$

Cl^- in surplus:

$$\begin{matrix} Cl^- & Cl^- & Cl^- & \\ Cl^- & [AgCl] & Cl^- + FI^- & \text{(yellow-green)} \\ Cl^- & Cl^- & Cl^- & \end{matrix}$$

Ag^+ in excess:

$$\begin{matrix} Ag^+ & Ag^+ & Ag^+ \\ Ag^+ & [AgCl] & Ag^+ \\ Ag^+ & Ag^+ & Ag^+ \end{matrix} + FI^- \xrightarrow{\text{adsorption}} \begin{matrix} Ag^+ & Ag^+ & Ag^+ \\ Ag^+ & [AgCl] & Ag^+FI^- \text{ (pink)} \\ Ag^+ & Ag^+ & Ag^+ \end{matrix}$$

2.3.2 The conditions for Fajans titration

(1) A large surface area: Consider adsorption occurs at the surface of the colloidal precipitate, the larger the surface area of the precipitate is, the greater the color change at the end point is. To prevent agglomerate of colloidal precipitate, a protective colloid, such as dextrin or starch, is often added to the more dilute titration solution before titrating, and a large number of neutral salt must be avoided.

(2) The adsorption ability: The adsorption ability of colloidal surface for indicator should moderately weak than for halide ions, otherwise the indicator will change color before the stoichiometric point. On the other hand, if the adsorption ability of colloidal surface for the indicator is too weak, the indicator cannot change color even after the stoichiometric point. The adsorption ability of AgX colloidal surface for common indicators and X^- is in the order of: I^- > Dimethyldiiodidefluorescein > Br^- > eosin > Cl^- > fluorescein. So, fluorescein is selected for the titration of Cl^- and eosin is selected for the titration of Br^-.

(3) Acidity: Most of the adsorption indicators are organic weak acids dissociated into an anionic ion in solution. To make sure the indicators negatively charged, the suitable acidity of solution should be $pK_a < pH < 10$. Acidity of the solution should be moderately lower when the dissociation constant (K_a) of the indicator is small, and Acidity of the solution should be moderately higher when K_a is large. Compare fluorescein and eosin, K_a of fluorescein is 10^{-7} and of eosin is 10^{-2}, so fluorescein is used at pH 7–10, but eosin is used at pH 2–10.

(4) The charge of indicator ion is opposite to that of standard solution ion. For example, Ag^+ can be titrated with positively charged fluorescein, but Cl^- can be titrated with negatively charged methyl violet.

(5) Keep away from bright light. Because of the susceptibility to the action of light, AgX decomposes into dark grey colored silver under light which makes the sensitivity of end-point detection is correspondingly decreased. For best results, the titrations should be carried out with minimum exposure to light.

The common adsorption indicators for applications are shown in Table 6-2.

Table 6-2 The common adsorption indicators for applications

Indicator	Analytes	Titrant	pH for titration
Fluorescein	Cl^-	Ag^+	7–10
Dichlorofluorescein	Cl^-	Ag^+	4–10
Eosin	Br^-, I^-, SCN^-	Ag^+	2–10
Methyl violet	SO_4^{2-}, Ag^+	Ba^{2+}, Cl^-	1.5–3.5
Dimethyldiiodidefluorescein	I^-	Ag^+	Neutral pH

2.3.3 Application

The Fajans method is mainly used for the determination of Cl^-, Br^-, I^-, SCN^-, and Ag^+, etc.

3. Standard solution and primary standard

3.1 Standard solution of $AgNO_3$

$AgNO_3$ is a primary standard and the usual titrant, the standard solution could be prepared directly and be calculated exactly with guarantee reagent (G.R.) of $AgNO_3$. But mostly, $AgNO_3$ solution is prepared indirectly with analytical reagent (A.R.) of $AgNO_3$ and is standardized with guarantee reagent (G.R.) of NaCl. NaCl, also a primary standard, has to be stored in a desiccator because of fairly hygroscopic from air or environment.

To prepare about 0.1mol/L $AgNO_3$ solution, about 17g of $AgNO_3$ is weighed, dissolved and diluted to about 1000ml using distilled water, then this solution is closed stored in a brown bottle or in the dark.

To standardize 0.1mol/L $AgNO_3$ with NaCl, 0.12–0.15g of NaCl (G.R.) is weighed accurately, and dissolve with about 50ml of distilled water in a conical flask. Then about 1ml of K_2CrO_4 indicator is added. This full shaken NaCl solution is titrated with $AgNO_3$ solution to be tasted and end point occurs when the color turns pink.

3.2 Standard solution of NH_4SCN

To prepare about 0.1mol/L NH_4SCN solution, about 8g of NH_4SCN is weighed, dissolved and diluted to about 1000ml using distilled water.

To standardize 0.1mol/L NH_4SCN with $AgNO_3$, 25.00ml of 0.1000mol/L $AgNO_3$ standard solution is measured accurately into a conical flask, and about 50ml of distilled water, 3ml of HNO_3 and 2ml of $NH_4Fe(SO_4)_2 \cdot 12H_2O$ is added in turn. This full shaken $AgNO_3$ solution is titrated with NH_4SCN solution to be tasted and end point occurs when the color turns light red with vigorous shaking. The concentration of NH_4SCN can be calculated by its volume been used.

PPT

Section 3 Application Examples

1. Determination of inorganic halide and inorganic hydrohalide

1.1 Determination of sodium chloride in sodium chloride injection

20.00ml of the injection to be tasted is measured accurately first. Then this injection is titrated with 0.1000mol/L $AgNO_3$ standard solution using K_2CrO_4 as an indicator. End point occurs when the color changes from yellow to yellow-green. (Concentration: g/100ml) (M_{NaCl} = 58.443g/mol)

$$\omega_{NaCl}(\%) = \frac{c_{AgNO_3} \times V_{AgNO_3} \times M_{NaCl}}{V_{NaCl} \times 1000} \times 100\%$$

Pharmacopoeia of the People's Republic of China (2015): Content of NaCl injection should be at 0.850%–0.950%.

1.2 Determination of ammonium chloride in white sal-ammoniac

1.2g of white sal-ammoniac to be tasted is weighed accurately, then dissolved, diluted to exactly 250.0ml with distilled water, shaken well in a volumetric flask.

25.00ml of supernatant liquid is measured accurately into a conical flask after standing some time. 25ml of distilled water, 3ml of HNO_3 and exactly 40.00ml of 0.1000mol/L $AgNO_3$ are added in turn. Then 3ml of nitrobenzene is added with vigorous shaking. 2ml of $NH_4Fe(SO_4)_2 \cdot 12H_2O$ is added into the solution too. The salmiac solution was titrated with 0.1000mol/L NH_4SCN standard solution and end point occurs when the color turn to red-brown.

$$\omega_{NH_4Cl}(\%) = \frac{[(c \cdot V)_{AgNO_3} - (c \cdot V)_{NH_4SCN}] \times M_{NH_4Cl}}{m_s \times \frac{25.00}{250.0} \times 1000} \times 100\%$$

1.3 Determination of ephedrine hydrochloride

15 tablets of ephedrine hydrochloride ($C_{10}H_{15}NO \cdot HCl$: 30mg or 25mg) to be tasted is weighed accurately and marked as m g. m_s g of comminuted powder ($C_{10}H_{15}NO \cdot HCl$: about 0.15g) is weighed accurately and this powder is dissolved with 25ml of distilled water with vigorous shaking, 2 drops of bromophenol blue is added, the titration with acetic acid stopes when color changes from violet to yellow-green. 10 drops of bromophenol blue and 5ml of 2% (g/ml) dextrin is added. The complex solution is titrated with 0.1mol/L $AgNO_3$ standard solution and end point occurs when the color turn to grey violet. 1ml of 0.1mol/L $AgNO_3$ standard solution is equivalent to 20.12mg of $C_{10}H_{15}NO \cdot HCl$.

$$\overline{\omega}\,(\text{mg/tablet}) = \frac{T \times V_{AgNO_3}}{m_s} \times \frac{m}{15}$$

$$C_{10}H_{15}NO \cdot HCl(\%) = \frac{\overline{\omega}}{\omega_s\,(\text{labeled amount})} \times 100\%$$

2. Pretreatment of organic halides

Because of the different combination of halogen in organic halide from inorganic halide, most of them cannot be titrated directly by argentometry and must be converted into inorganic halogens through proper treatment. The common treatment methods include alkaline hydrolysis, sodium carbonate fusion and oxygen flask combustion method, etc.

重点小结

沉淀滴定法滴定曲线的特点：①pX与pAg两条曲线以化学计量点对称；②突跃范围的大小，取决于沉淀的溶度积常数K_{sp}和溶液的浓度c；③当溶液中Cl、Br和I共存时可分步沉淀，在滴定曲线上显示三个突跃。

银量法常用三种指示终点方法：①铬酸钾指示剂法（莫尔法）；②铁铵钒指示剂法（佛尔哈德法），包括直接滴定法和返滴定法；③吸附指示剂法（法扬司法）。

指示终点方法		化学反应式	
莫尔法		终点前：$Ag^+ + Cl^- \rightleftharpoons AgCl\downarrow$ 终点时：$2Ag^+ + CrO_4^{2-} \rightleftharpoons Ag_2CrO_4\downarrow$	(白色) (砖红色)
佛尔哈德法	直接滴定	终点前：$Ag^+ + SCN^- \rightleftharpoons AgSCN\downarrow$ 终点时：$Fe^{3+} + SCN^- \rightleftharpoons [Fe(SCN)]^{2+}\downarrow$	(白色) (红色)
	返滴定	终点前：Ag^+(准确过量)$+X^- \rightleftharpoons AgX\downarrow$ Ag^+(剩余)$+SCN^- \rightleftharpoons AgSCN\downarrow$ 终点时：$Fe^{3+} + SCN^- \rightleftharpoons [Fe(SCN)]^{2+}\downarrow$	 (白色) (红色)
法扬司法		终点前：Cl^-(剩余) $(AgCl)\cdot Cl^- + FI^-$ 终点时：Ag^+(稍过量) $(AgCl)\cdot Ag^+\cdot FI^-$	(黄绿色) (淡红色)

三种指示终点方法的操作条件比较如下：

指示终点方法		被测离子	滴定剂	指示剂	滴定条件
莫尔法		Cl^- 和 Br^-	$AgNO_3$	铬酸钾 K_2CrO_4	中性或弱碱性，pH 6.5-10.5； NH_4^+ 存在时，pH 6.5-7.2
佛尔哈德法	直接滴定	Ag^+	NH_4SCN	铁铵矾 $NH_4Fe(SO_4)_2\cdot 12H_2O$	HNO_3 介质，酸性
	返滴定	Cl^-、Br^-、I^-、SCN^- 等	$AgNO_3$、NH_4SCN		
法扬司法		Cl^-、Br^-、I^-、SCN^- 和 Ag^+	$AgNO_3$	荧光黄等 吸附指示剂	$pK_a < pH < 10$

题库

目 标 检 测

1. Identification:

(1) Precipitation titration

(2) Argentometry

(3) Adsorption indicator

2. What will the result about end point error be, it is positive error, negative error or no influence by using Mohr method?

(1) Determine chloride (Cl^-) at pH 4.

(2) Determine chloride (Cl^-) in a NH_4Cl solution at pH 7.

(3) Determine chloride (Cl^-) in a mixture of NaCl and Na_2CO_3.

(4) Determine iodide (I^-) or thiocyanate (SCN^-).

3. What will the result about end point error be, it is positive error, negative error or no influence by using Volhard method?

(1) Determine chloride (Cl^-) without filtration of AgCl or addition of nitrobenzene.

(2) Determine iodide (Br^-) without filtration of AgBr or addition of nitrobenzene.

(3) Determine iodide (I^-), $NH_4Fe(SO_4)_2 \cdot 12H_2O$ indicator was added before the excess $AgNO_3$ standard solution.

(4) Determine chloride (Cl^-) in a mixture of NaCl and Na_2CO_3.

4. What will the result about end point error be, it is positive error, negative error or no influence by using Fajans method?

(1) Determine chloride (Cl^-) with eosin indicator.

(2) Determine bromide (Br^-) with eosin indicator.

(3) Determine iodide (I^-) with fluorescein indicator.

(4) Determine chloride (Cl^-) in a mixture of NaCl and Na_2SO_4.

5. Briefly describe the experimental design for the determination of KCl by using Mohr method, please list the titrant, the indicator and the major titration conditions.

6. Briefly describe the experimental design for the determination of KCl by using Volhard method, please list the titrant, the indicator and the major titration conditions.

7. Briefly describe the experimental design for the determination of KCl by using Fajans method, please list the titrant, the indicator and the major titration conditions.

8. Select a suitable method of argentometry to determine the following sample, please briefly state the reasons.

(1) $BaCl_2$

(2) $FeCl_3$

(3) NaCl and Na_3PO_4

(4) NaCl and Na_2CO_3

(5) NH_4Cl

(6) KSCN

(7) NaBr

(8) KI

9. A halite sample contains NaCl. A 0.1960g of such halite sample is dissolved in water and mixed with 30.00ml of 0.1220mol/L $AgNO_3$, using $NH_4Fe(SO_4)_2 \cdot 12H_2O$ indicator, excess $AgNO_3$ is then titrated with 6.50ml of 0.1020mol/L NH_4SCN standard solution. Please calculate percentage of NaCl in this halite sample. [*M* (g/mol): NaCl: 58.443]

10. A mixture sample contains KCl and KBr. Using gravimetry, a 0.5000g of this mixture sample can form A 0.4240g of silver salt; using argentometry, A 0.5000g of same sample is titrated with 22.40ml of 0.1050mol/L $AgNO_3$. Please calculate percentage of KCl and KBr. in this mixture sample. [*M* (g/mol): KCl: 74.551, KBr: 119.00, AgCl: 143.32, AgBr: 187.77]

11. A pesticide sample contains arsenic salt. A 0.2000g of thus pesticides is oxidized in HNO_3 to be H_3AsO_4. Then the resulted solution is adjusted to a neutral pH and H_3AsO_4 is precipitated as Ag_3AsO_4 by $AgNO_3$. Next, Ag_3AsO_4 is filtered, washed, and dissolved in diluted HNO_3. Last, using $NH_4Fe(SO_4)_2 \cdot 12H_2O$ indicator, the dissolved solution is titrated with 25.86ml of 0.1260mol/L NH_4SCN. Please calculate percentage of As_2O_3 in this mixture sample. [*M* (g/mol): As_2O_3: 197.84]

12. A 0.3012g of HgS sample, mixed with 10ml HNO_3 and 1.5g of K_2SO_4, is dissolved by heating, then 50ml of water is added at room temperature. 1% (g/ml) of $KMnO_4$ is added until the

color of thus solotion becomes pink. 2% (g/ml) of $FeSO_4$ is dripped until pink is just disappeared. 2ml $NH_4Fe(SO_4)_2 \cdot 12H_2O$ is added and the resulted solution is titrated with 19.95ml of 0.1260mol/L NH_4SCN standard solution. Please calculate percentage of HgS in this sample, and determine whether the product is qualified. [1ml of 0.1000mol/L NH_4SCN standard solution can titrate just 11.63mg of HgS] [standard: HgS≥96.0%]

Chapter 7 Complexometric Titration

学习目标

知识要求

1. **掌握** 配位滴定法的基本概念和基本原理；副反应对EDTA及金属离子配合物稳定性的影响；稳定常数和条件稳定常数的定义及表达形式；金属指示剂指示终点的作用原理及适用条件。

2. **熟悉** EDTA及其与金属离子形成配合物的特点；配位滴定曲线及其影响滴定突跃的因素。

3. **了解** 配位滴定的滴定方式；EDTA在分析测定中的应用；配位滴定的应用。

能力要求

通过学习配位滴定法基础理论和相关知识，对配位滴定法的基本概念和原理有系统的认识，并且可以处理相关问题。

PPT

Section 1 Basic Principles of Complexometric Titration

1. Introduction to complexometric titration

Many metal ions form slightly dissociated complexes with various ligands (complexing agents). The analytical chemist makes judicious use of complexes to mask undesired reactions. The formation of complexes can also serve as the basis of accurate and convenient titrations for metal ions in which the titrant is a complexing agent. Complexometric titrations (配位滴定法) are useful for determining a large number of metals. Selectivity can be achieved by appropriate use of masking agents (掩蔽剂) (addition of other complexing agents that react with interfering metal ions, but not with the metal of interest) and by pH control, since most complexing agents are weak acids or weak bases whose equilibria are influenced by the pH. In this chapter, we discuss metal ions, their equilibria, and the influence of pH on the equilibria. We describe titrations of metal ions with the very useful complexing agent (配位剂) EDTA, the factors that affect them, and indicators for the titrations. The EDTA titration of calcium plus magnesium is commonly used to determine water hardness (水硬度). In the food industry, calcium is determined in cornflakes. In the plating industry, nickel is determined in plating solutions by complexometric titration, also called chelometric titration (螯合滴定法), and in the metals

industry in etching solutions. In the pharmaceutical industry, aluminum hydroxide in liquid antacids is determined by similar titrations. Nearly all metals can be accurately determined by complexometric titrations. Complexing reactions are useful for gravimetry, spectrophotometry, fluorometry and masking interfering ions.

1.1 Complexation reaction

Most metal ions react with electron-pair donors to form coordination compounds (配位化合物) or complexes (配合物). The donor species, or ligand, must have at least one pair of unshared electrons available for bond formation.

The number of covalent bonds that a cation tends to form with electron donors is its coordination number. Typical values for coordination numbers are two, four and six. The species formed as a result of coordination can be electrically positive, neutral, or negative. For example, copper (II), which has a coordination number of four, forms a cationic ammine complex, $Cu(NH_3)_4^{2+}$; a neutral complex with glycine, $Cu(NH_2CH_2COO)_2$; and an anionic complex with chloride ion, $CuCl_4^{2-}$.

Complexation reactions (配位反应) are widely used in analytical chemistry. One of the earliest uses of these reactions was for titrating cations, a major topic of this chapter. In addition, many complexes are colored or absorb ultraviolet radiation; the formation of these complexes is often the basis for spectrophotometric determinations. Some complexes are sparingly soluble and can be used in gravimetric analysis or for precipitation titrations. Complexes are also widely used for extracting cations from one solvent to another and for dissolving insoluble precipitates. The most useful complex forming reagents are organic compounds containing several electron-donor groups that form multiple covalent bonds with metal ions. Inorganic complexing agents are also used to control solubility, form colored species, or form precipitates.

1.2 Complexometric titration

Complexometric titrations are useful for determining a large number of metals. The earliest titrimetric applications involving metal-ligand complexation were the determinations of cyanide and chloride using, respectively, Ag^+ and Hg^{2+} as titrants. Both methods were developed by Justus Liebig (1803–1873) in the 1850s. The use of a monodentate ligand, such as Cl^- and CN^-, however, limited the utility of complexation titrations to those metals that formed only a single stable complex, such as $Ag(CN)_2^-$ and $HgCl_2$. Other potential metal-ligand complexes, such as CdI_4^{2-}, were not analytically useful because the stepwise formation of a series of metal-ligand complexes (CdI^+, CdI_2, CdI_3^-, and CdI_4^{2-}) resulted in a poorly defined end point.

Titrations based on complex formation, sometimes called complexometric titrations, have been used for more than a century. The utility of complexation titrations improved following the introduction by Schwarzenbach, in 1945, of aminocarboxylic acids as multidentate ligands capable of forming stable 1 : 1 complex with metal ions. The most widely used of these new ligands was ethylenediaminetetraacetic acid (乙二胺四乙酸), EDTA, which forms strong 1 : 1 complex with many metal ions. The first use of EDTA as a titrant occurred in 1946, when Schwarzenbach introduced metallochromic dyes as visual indicators for signaling the end point of a complexation titration.

2. Properties of EDTA and its complexes

EDTA is a merciful abbreviation for ethylenediaminetetraacetic acid, a compound that forms strong 1 : 1 complexes with most metal ions and finds wide use in quantitative analysis (定量分析). The structure of EDTA is shown in Figure 7-1.

Figure 7-1 The structure of EDTA

EDTA, which is a Lewis acid (路易斯酸), has six binding sites (the four carboxylate groups and the two amino groups), providing six pairs of electrons. The resulting metal–ligand complex, in which EDTA forms a cage-like structure around the metal ion (see Figure 7-2), is very stable. The actual number of coordination sites depends on the size of the metal ion.

Figure 7-2 EDTA forms strong 1 : 1 complexes with most metal ions

EDTA plays a larger role as a strong metal-binding agent in industrial processes and in products such as detergents, cleaning agents, and food additives that prevent metal-catalyzed oxidation of food. EDTA is an emerging player in environmental chemistry. For example, the majority of nickel discharged into San Francisco Bay and a significant fraction of the iron, lead, copper, and zinc are EDTA complexes that pass unscathed through wastewater treatment plants.

The free acid H_4Y and the dihydrate of the sodium salt, $Na_2H_2Y \cdot 2H_2O$, are commercially available in reagent quality. The free acid can serve as a primary standard after it has been dried for several hours at 130°C to 145°C. However, the free acid is not very soluble in water and must be dissolved in a small amount of base for complete solution.

More commonly, the dihydrate, $Na_2H_2Y \cdot 2H_2O$, is used to prepare standard solutions (标准溶液). Under normal atmospheric conditions, the dihydrate contains 0.3% moisture in excess of the stoichiometric water of hydration. For all but the most exacting work, this excess is sufficiently reproducible to permit use of a corrected mass of the salt in the direct preparation of a standard solution. If necessary, the pure dihydrate can be prepared by drying at 80°C for several days in an atmosphere of 50% relative humidity (相对湿度). Alternatively, an approximate concentration can be prepared and then standardized against primary standard $CaCO_3$.

Several compounds that are chemically related to EDTA have also been investigated. Since these do not seem to offer significant advantages, we shall limit our discussion here to the properties and applications of EDTA.

2.1 Dissociation equilibrium of EDTA in aqueous solution

EDTA is a hexaprotic system, designated H_6Y^{2+}. The highlighted, acidic hydrogen atoms are the ones that are lost upon metal-complex formation.

$$H_6Y^{2+} \rightleftharpoons H^+ + H_5Y^+ \qquad pK_{a1} = 0.0 \qquad (7\text{-}1)$$

$$H_5Y^+ \rightleftharpoons H^+ + H_4Y \qquad pK_{a2} = 1.5 \qquad (7\text{-}2)$$

$$H_4Y \rightleftharpoons H^+ + H_3Y^- \qquad pK_{a3} = 2.0 \qquad (7\text{-}3)$$

$$H_3Y^- \rightleftharpoons H^+ + H_2Y^{2-} \qquad pK_{a4} = 2.66 \qquad (7\text{-}4)$$

$$H_2Y^{2-} \rightleftharpoons H^+ + HY^{3-} \qquad pK_{a5} = 6.16 \qquad (7\text{-}5)$$

$$HY^{3-} \rightleftharpoons H^+ + Y^{4-} \qquad pK_{a6} = 10.24 \qquad (7\text{-}6)$$

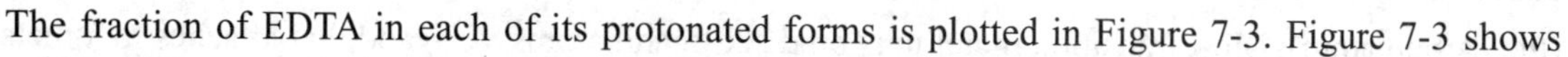

The fraction of EDTA in each of its protonated forms is plotted in Figure 7-3. Figure 7-3 shows the fraction of species plot that is obtained when we do this, where Y^{4-} becomes the principle form only at high pH values. Since the anion Y^{4-} is the ligand species in complex formation, the complexation equilibria are affected markedly by the pH. H_4Y has a very low solubility in water, and so the disodium salt $Na_2H_2Y \cdot 2H_2O$ is generally used, in which two of the acid groups are neutralized. This salt dissociates in solution to give predominantly H_2Y^{2-}; the pH of the solution is approximately 4 to 5 (theoretically 4.4 from $[H^+]=K_{a2}K_{a3}$).

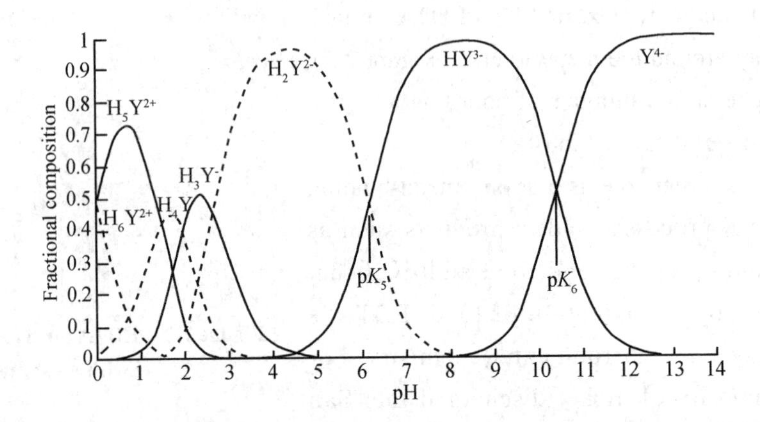

Figure 7-3 Fractional composition diagram for EDTA

The following ladder diagram (Figure 7-4) shows the distribution of EDTA species at varied pH. This ladder diagram was contributed by Professor Galina Talanova, Howard University. As noted above, the nitrogens on EDTA can protonate in very acidic solution, and when considering this, we can write six K_a values, in which case the pK_1–pK_6 values in the ladder diagram correspond to K_{a1}–K_{a6} in Equations 7-1 through 7-6 (note the protonated nitrogens occur around pH 0 and 1.5):

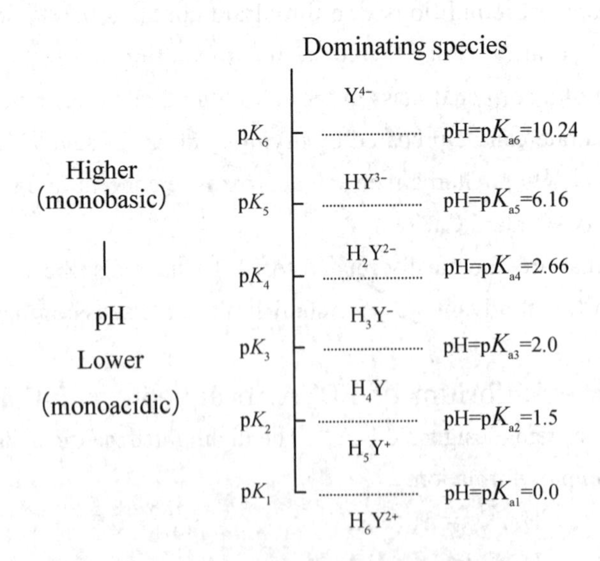

Figure 7-4 Distribution diagram of EDTA species at varied pH

2.2 Analytical properties of metal-EDTA complexes

Metal ions are Lewis acids, accepting electron pairs from electron-donating ligands that are Lewis bases. Cyanide is called a monodentate ligand because it binds to a metal ion through only one atom (the carbon atom). Most transition metal ions bind six ligand atoms. A ligand that attaches to a metal ion through more than one ligand atom is said to be multidentate ("many toothed"), or a chelating ligand (pronounced KEE-late-ing).

3. The dissociation equilibrium of the complex in solution

3.1 Stability of complexes formed by EDTA with metal ions

The equilibrium constant for the reaction of a metal with a ligand is called the formation constant (稳定常数) K_{MY}. The formation constant is also called the stability constant K_s, or $K_{stab.}$

$$M^{n+} + Y^{4-} \rightleftharpoons MY^{n-4} \quad K_{MY} = \frac{[MY^{n-4}]}{[M^{n+}][Y^{4-}]} \tag{7-7}$$

Note that K_{MY} for EDTA is defined in terms of the species reacting with the metal ion. The equilibrium constant could have been defined for any of the other six forms of EDTA in the solution. Equation 7-7 should not be interpreted to mean that only reacts with metal ions. Table 7-1 shows that formation constants for most EDTA complexes are quite large and tend to be larger for more positively charged cations.

Table 7-1 Formation constants for metal-EDTA complexes

Ion	lg K_f	Ion	lg K_f	Ion	lg K_f
Li^+	2.95	V^{3+}	25.9[a]	Tl^{3+}	35.3
Na^+	1.86	Cr^{3+}	23.4[a]	Bi^{3+}	27.8[a]
K^+	0.8	Mn^{3+}	25.2	Ce^{3+}	15.93
Be^{2+}	9.7	Fe^{3+}	25.1	Pr^{3+}	16.30
Mg^{2+}	8.79	Co^{3+}	41.4	Nd^{3+}	16.51
Ca^{2+}	10.65	Zr^{4+}	29.3	Pm^{3+}	16.9
Sr^{2+}	8.72	Hf^{4+}	29.5	Sm^{3+}	17.06
Ba^{2+}	7.88	VO^{2+}	18.7	Eu^{3+}	17.25
Ra^{2+}	7.4	VO_2^+	15.5	Gd^{3+}	17.35
Sc^{3+}	23.1[a]	Ag^+	7.20	Tb^{3+}	17.87
Y^{3+}	18.08	Tl^+	6.41	Dy^{3+}	18.30
La^{3+}	15.36	Pd^{2+}	25.6[a]	Ho^{3+}	18.56
V^{2+}	12.7[a]	Zn^{2+}	16.5	Er^{3+}	18.89
Cr^{2+}	13.6[a]	Cd^{2+}	16.5	Tm^{3+}	19.32
Mn^{2+}	13.89	Hg^{2+}	21.5	Yb^{3+}	19.49
Fe^{2+}	14.30	Sn^{2+}	18.3[b]	Lu^{3+}	19.74
Co^{2+}	16.45	Pb^{2+}	18.0	Th^{4+}	23.2
Ni^{2+}	18.4	Al^{3+}	16.4	U^{4+}	25.7
Cu^{2+}	18.78	Ga^{3+}	21.7		

continued

Ion	lg K_f	Ion	lg K_f	Ion	lg K_f
Ti^{3+}	21.3	In^{3+}	24.9		

NOTE: The stability constant is the equilibrium constant for the reaction $M^{n+} + Y^{4-} \rightleftharpoons MY^{n-4}$. Values in table apply at 25°C and ionic strength 0.1mol/L unless otherwise indicated.

a. 20°C, ionic strength=0.1mol/L. b. 20°C, ionic strength=1mol/L.

SOURCE: Martell A E, Smith R M, Motekaitis R J. NIST critically selected stability constants of metal complexes[M]. NIST Standard Reference Database 46, Gaithersburg, MD, 2001.

【Example 7-1】

A divalent metal M^{2+} reacts with a ligand L to form a 1 : 1 complex:

$$M^{2+}+L \rightleftharpoons ML^{2+} \quad K_{MY}=\frac{[ML^{2+}]}{[M^{2+}][L]}$$

Calculate the concentration of M^{2+} in a solution prepared by mixing equal volumes of 0.20mol/L M^{2+} and 0.20mol/L L. K_{ML}=1.0×10^8.

Solution: We have added stoichiometrically equal amounts of M^{2+} and L. The complex is sufficiently strong that their reaction is virtually complete. Since we added equal volumes, the initial concentrations were halved by dilution. Let x represent $[M^{2+}]$. At equilibrium, we have

$$\begin{array}{ccc} M^{2+} + & L \rightleftharpoons & ML^{2+} \\ x & x & 0.10-x\approx 0.10 \end{array}$$

Essentially, all the M^{2+} (original concentration 0.20mol/L) was converted to an equal amount of ML^{2+}, with only a small amount of uncomplexed metal remaining. Substituting into the K_{ML} expression,

$$\frac{0.10}{x\cdot x}=1.0\times 10^{8}$$

$$x=[M^{2+}]=3.2\times 10^{-5}\text{mol/L}$$

You can also solve the equation $(0.10-x)/x^2=K_{ML}$ by Goal Seek in Excel. This will apply even when K_{ML} is not so high and you cannot really assume that x is so small that 0.10−x can be approximated as 0.10.

3.2 Factors affecting the stability of EDTA complexes

3.2.1 Side reactions of coordination agent (Y)

Side reactions of coordination agent measured by side reaction coefficient, α. The fraction of all free EDTA in the form Y^{4-} is called $\delta_{Y^{4-}}$. It is reciprocal to the side reaction coefficient.

$$\delta_{Y^{4-}}=\frac{[Y^{4-}]}{[H_6Y^{2+}]+[H_5Y^{+}]+[H_4Y]+[H_3Y^{-}]+[H_2Y^{2-}]+[HY^{3-}]+[Y^{4-}]}$$

$$\delta_{Y^{4-}}=\frac{[Y^{4-}]}{[EDTA]} \tag{7-8}$$

Where [EDTA] is the total concentration of all free EDTA species in the solution. By "free," we mean EDTA not complexed to metal ions. Following the derivation in Section 5.2 , we can show that is given by

$$\delta_{Y^{4-}}=\frac{[Y^{4-}]}{[EDTA]}=\frac{1}{\alpha_{Y^{4-}}}$$

$$= \frac{K_1K_2K_3K_4K_5K_6}{[H^+]^6 + [H^+]^5K_1 + [H^+]^4K_1K_2 + \cdots + [H^+]K_1K_2K_3K_4K_5 + K_1K_2K_3K_4K_5K_6}$$

Table 7-2 gives values for as a function of pH.

Table 7-2 Values of $\delta_{Y^{4-}}$ for EDTA at 20°C and μ=0.10mol/L

pH	$\delta_{Y^{4-}}$	pH	$\delta_{Y^{4-}}$
0	1.3×10^{-23}	8	4.2×10^{-3}
1	1.4×10^{-18}	9	0.041
2	12.6×10^{-14}	10	0.30
3	2.1×10^{-11}	11	0.81
4	3.0×10^{-9}	12	0.98
5	2.9×10^{-7}	13	1.00
6	1.8×10^{-5}	14	1.00
7	3.8×10^{-4}		

【Example 7-2】

The fraction of all free EDTA in the form Y^{4-} is called $\delta_{Y^{4-}}$. At pH 6.00 and a formal concentration of 0.10mol/L, the composition of an EDTA solution is

$[H_6Y^{2+}]$=8.9×10^{-20}mol/L; $[H_5Y^+]$=8.9×10^{-14}mol/L; $[H_4Y]$=2.8×10^{-7}mol/L;
$[H_3Y^-]$=2.8×10^{-5}mol/L; $[H_2Y^{2-}]$=0.057mol/L; $[HY^{3-}]$=0.043mol/L;
$[Y^{4-}]$=1.8×10^{-6}mol/L. Find $\delta_{Y^{4-}}$.

Solution: $\delta_{Y^{4-}}$ is the fraction in the form Y^{4-}:

$$\delta_{Y^{4-}} = \frac{[Y^{4-}]}{[H_6Y^{2+}] + [H_5Y^+] + [H_4Y] + [H_3Y^-] + [H_2Y^{2-}] + [HY^{3-}] + [Y^{4-}]}$$

$$= \frac{1.8\times10^{-6}}{8.9\times10^{-20} + 8.9\times10^{-14} + 2.8\times10^{-7} + 2.8\times10^{-5} + 0.057 + 0.043 + 1.8\times10^{-6}}$$

$$=1.8\times10^{-5}$$

【Example 7-3】

Calculate the molar Y^{4-} concentration in a 0.0200mol/L EDTA solution buffered to a pH of 10.00.

Solution: At pH 10.00, $\delta_{Y^{4-}}$ is 0.30 (see Table 7-2). Thus,

$$[Y^{4-}] = \delta_{Y^{4-}} c_{EDTA} = 0.30\times0.0200 = 6.00\times10^{-3}\text{mol/L}$$

3.2.2 Side reactions of metal ions M

Complexation reactions involve a metal-ion M reacting with a ligand L to form a complex ML, as shown in Equation 7-9:

$$M + L \rightleftharpoons ML \qquad K_1 = \frac{[ML]}{[M][L]} \tag{7-9}$$

Where we have omitted the charges on the ions in order to be general. Complexation reactions occur in a stepwise fashion and the reaction above is often followed by additional reactions:

$$ML + L \rightleftharpoons ML_2 \qquad K_2 = \frac{[ML_2]}{[ML][L]} \tag{7-10}$$

$$ML_2 + L \rightleftharpoons ML_3 \qquad K_3 = \frac{[ML_3]}{[ML_2][L]} \tag{7-11}$$

$$\vdots \qquad \vdots$$

$$ML_{n-1} + L \rightleftharpoons ML_n \qquad K_n = \frac{[ML_n]}{[ML_{n-1}][L]} \tag{7-12}$$

Unidentate ligands invariably add in a series of steps as shown above. With multidentate ligands, the maximum coordination number of the cation may be satisfied with only one or a few added ligands. For example, Cu(Ⅱ), with a maximum coordination number of 4, can form complexes with ammonia that have the formula as $Cu(NH_3)^{2+}$, $Cu(NH_3)_2^{2+}$, $Cu(NH_3)_3^{2+}$, and $Cu(NH_3)_4^{2+}$. With the bidentate ligand glycine (gly), the only complexes that form are $Cu(gly)^{2+}$ and $Cu(gly)_2^{2+}$.

The equilibrium constants for complex formation reactions are generally written as formation constants (形成常数). Thus, each of the Reactions 7-9 through 7-12 is associated with a stepwise formation constant K_1 through K_n. We can also write the equilibria as the sum of individual steps. These have overall formation constants designated by the symbol β_n, cumulative stability constant (累积稳定常数). Therefore,

$$M + L \rightleftharpoons ML \qquad \beta_1 = K_1 = \frac{[ML]}{[M][L]} \tag{7-13}$$

$$M + 2L \rightleftharpoons ML_2 \qquad \beta_2 = K_1 \cdot K_2 = \frac{[ML_2]}{[M][L]^2} \tag{7-14}$$

$$M + 3L \rightleftharpoons ML_3 \qquad \beta_3 = K_1 \cdot K_2 \cdot K_3 = \frac{[ML_3]}{[M][L]^3} \tag{7-15}$$

$$\vdots \qquad \vdots$$

$$M + nL \rightleftharpoons ML_n \qquad \beta_n = K_1 \cdot K_2 \cdots K_n = \frac{[ML_n]}{[M][L]^n} \tag{7-16}$$

Except for the first step, the overall formation constants (总形成常数) are products of the stepwise formation constants for the individual steps leading to the product.

For a given species like the free metal M, we can calculate an alpha value, which is the fraction of the total metal concentration in that form. δ_M is the fraction of the total metal present at equilibrium in the free metal form. It is reciprocal to the side reaction coefficient (α_M)

$$\delta_M = \frac{[M]}{c_M} = \frac{1}{\alpha_M} \tag{7-17}$$

The total metal concentration c_M can be written

$$c_M = [M] + [ML] + [ML_2] + \ldots + [ML_n]$$

From the overall formation constants (Equations 7-13 through 7-16), the concentrations of the complexes can be expressed in terms of the free metal concentration [M] to give

$$c_M = [M] + \beta_1[M][L] + \beta_2[M][L]^2 + \cdots + \beta_n[M][L]^n \tag{7-18}$$

Then δ_M can be found as

$$\delta_M = \frac{[M]}{c_M} = \frac{[M]}{[M] + \beta_1[M][L] + \beta_2[M][L]^2 + \cdots + \beta_n[M][L]^n}$$

$$= \frac{1}{1+\beta_1[\mathrm{L}]+\beta_2[\mathrm{L}]^2+\cdots+\beta_n[\mathrm{L}]^n} \tag{7-19}$$

$$\alpha_{\mathrm{M}} = \frac{1}{\delta_{\mathrm{M}}} = 1+\beta_1[\mathrm{L}]+\beta_2[\mathrm{L}]^2+\cdots+\beta_n[\mathrm{L}]^n$$

【Example 7-4】

Zn^{2+} and NH_3 form the complexes $Zn(NH_3)^{2+}$, $Zn(NH_3)_2^{2+}$, $Zn(NH_3)_3^{2+}$, and $Zn(NH_3)_4^{2+}$. If the concentration of free, unprotonated NH_3 is 0.10mol/L, find the fraction of zinc in the form Zn^{2+}. (At any pH, there will also be some NH_4^+ in equilibrium with NH_3.)

Solution: The formation constants for the complexes: $Zn(NH_3)^{2+}$(β_1=$10^{2.18}$), $Zn(NH_3)_2^{2+}$(β_2=$10^{4.43}$), $Zn(NH_3)_3^{2+}$(β_3=$10^{6.74}$), and $Zn(NH_3)_4^{2+}$(β_4=$10^{8.70}$). The appropriate form of Equation 7-19 is

$$\delta_{\mathrm{Zn}^{2+}} = \frac{1}{1+\beta_1[\mathrm{L}]+\beta_2[\mathrm{L}]^2+\beta_3[\mathrm{L}]^3+\beta_4[\mathrm{L}]^4} \tag{7-20}$$

Equation 7-20 gives the fraction of zinc in the form Zn^{2+}. Putting in [L] = 0.10mol/L and the four values of β_i gives $\delta_{\mathrm{Zn}^{2+}}$=$1.8\times10^{-5}$, which means there is very little free Zn^{2+} in the presence of 0.10mol/L NH_3.

3.2.3 The conditional stability constant of EDTA complex

The formation constant describes the reaction between and a metal ion. As you see in Figure 7-3, most EDTA is not below pH 10.37. The species and so on, predominate at lower pH. From the definition we can express the concentration of as

$$[\mathrm{Y}^{4-}] = \delta_{\mathrm{Y}^{4-}}[\mathrm{EDTA}]$$

where [EDTA] is the total concentration of all EDTA species not bound to metal ion.

The formation constant can now be rewritten as

$$K_{\mathrm{MY}} = \frac{[\mathrm{MY}^{n-4}]}{[\mathrm{M}^{n+}][\mathrm{Y}^{4-}]} = \frac{[\mathrm{MY}^{n-4}]}{[\mathrm{M}^{n+}]\delta_{\mathrm{Y}^{4-}}[\mathrm{EDTA}]}$$

If the pH is fifixed by a buffer, then $\delta_{\mathrm{Y}^{4-}}$ is a constant that can be combined with K_{MY}.

Conditional formation constant: $$K'_{\mathrm{MY}} = \delta_{\mathrm{Y}^{4-}}K_{\mathrm{MY}} = \frac{K_{\mathrm{MY}}}{\alpha_{\mathrm{Y}^{4-}}} = \frac{[\mathrm{MY}^{n-4}]}{[\mathrm{M}^{n+}][\mathrm{EDTA}]} \tag{7-21}$$

The number K'_{MY} is called the conditional formation constant (条件稳定常数), or the effective formation constant.

It describes the formation of MY^{n-4} at any particular pH. After we learn to use Equation 7-21, we will modify it to allow for the possibility that not all metal ion is in the form M^{n+}.

The conditional formation constant allows us to look at EDTA complex formation as if the uncomplexed EDTA were all in one form:

$$\mathrm{M}^{n+} + \mathrm{EDTA} \rightleftharpoons \mathrm{MY}^{n-4}$$

$$K'_{\mathrm{MY}} = \delta_{\mathrm{Y}^{4-}}K_{\mathrm{MY}} = \frac{K_{\mathrm{MY}}}{\alpha_{\mathrm{Y}^{4-}}} \tag{7-22}$$

Equation 7-22 may be expressed in the logarithm form

$$\lg K'_{\mathrm{MY}} = \lg K_{\mathrm{MY}} - \lg\alpha_{\mathrm{Y}^{4-}} \tag{7-23}$$

At any given pH, we can find and evaluate K'_{MY}.

【Example 7-5】

Calculate the values of $\lg K'_{\mathrm{ZnY}}$ at pH=2.0 and pH=5.0.

Solution: from Table 7-1, $\lg K_{\mathrm{ZnY}}$=16.5.

(1) At pH=2.0, $\lg\alpha_{\mathrm{Y}^{4-}}$=$\lg\alpha_{\mathrm{Y(H)}}$=13.5,

Thus

$$\lg K'_{ZnY} = \lg K_{ZnY} - \lg \alpha_{Y^{4-}}$$
$$= 16.5 - 13.5 = 3.0$$

(2) At pH=5.0, $\lg\alpha_{Y^{4-}}=\lg\alpha_{Y(H)}=6.6$,

Thus

$$\lg K'_{ZnY} = \lg K_{ZnY} - \lg \alpha_{Y^{4-}}$$
$$= 16.5 - 6.6 = 9.9$$

It may be seen from the calculation that in a solution of pH = 2.0, the EDTA chelate of Zn^{2+} is unstable ($\lg K'_{ZnY}$ is only 3.0); when acidity of the solution decreases to pH 5.0, however, the chelate becomes more stable ($\lg K'_{ZnY}$ is 9.9).

Generally speaking, the stability of EDTA chelate increases with increasing the pH of the solution, and this may be beneficial for the complexometric titration of the metal ion. But if the pH is too high, the metal ion may hydrolyze, forming hydroxide. Taking these two factors into consideration, we can find a minimum pH value, at which a metal ion can be effectively titrated.

This minimum pH value for a given metal ion depends on the tolerable error and the required pM (–lg[M]) change in the end point detection. The former is usually ±0.1%, and the ΔpM for the visual end point detection is about ±0.2. Thus the criterion for effective titration of a single metal can be derived as

$$\lg cK_{MY} \geqslant 6 \tag{7-24}$$

Substituting Equation 7-23 into Equation 7-24 and rearranging the equation, we have

$$\lg\alpha_{Y(H)} \leqslant \lg K_{MY} + \lg c - 6 \tag{7-25}$$

Then, the minimum pH value can be obtained.

Figure 7-5 shows the minimum pH at which different metal ions can be titrated with EDTA ([M]=0.01000mol/L).

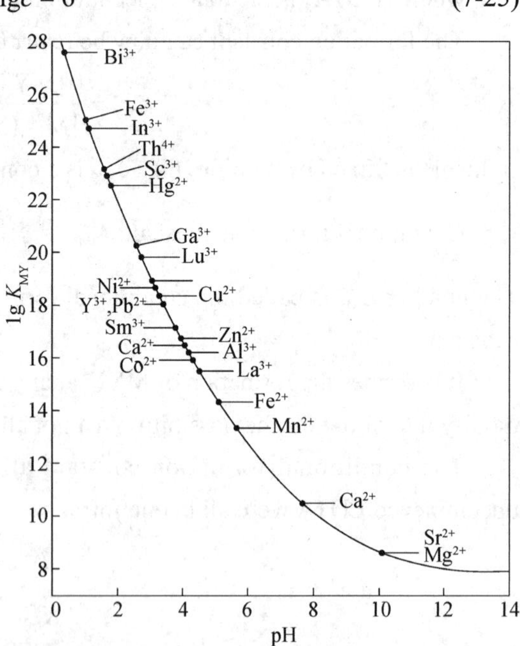

Figure 7-5 Minimum pH for effective titration of various metal ions by EDTA

4. Titration curves

Now that we know something about EDTA's chemical properties, we are ready to evaluate its utility as a titrant for the analysis of metal ions. To do so we need to know the shape of a complexometric EDTA titration curve. In Section 5.4 we saw that an acid–base titration curve shows the change in pH following the addition of titrant. The analogous result for a titration with EDTA shows the change in pM, where M is the metal ion, as a function of the volume of EDTA. In this section we learn how to calculate the titration curve.

Let's see how to reproduce the EDTA titration curves in Figure 7-6 by using one equation that applies to the entire titration. Because the reactions are carried out at fixed pH, the equilibria and mass balances are sufficient to solve for all unknowns. Consider the titration of metal ion M (concentration=c_M, initial volume=V_M) with a solution of ligand L (concentration=c_L, volume added=V_L) to form a 1 : 1 complex:

$$M + Y \rightleftharpoons MY$$

$$K_{MY}=\frac{[MY]}{[M][Y]}\Rightarrow[MY]=K_{MY}[M][Y] \tag{7-26}$$

The mass balances for metal and ligand are

Mass balance for M: $[M]+[MY]=\dfrac{c_M V_M}{V_M+V_Y}$

Mass balance for Y: $[Y]+[MY]=\dfrac{c_Y V_Y}{V_M+V_Y}$

Substituting $K_{MY}[M][Y]$ (from Equation 7-26) for [MY] in the mass balances gives

$$[M](1+K_{MY}[Y])=\frac{c_M V_M}{V_M+V_Y} \tag{7-27}$$

$$[Y](1+K_{MY}[M])=\frac{c_Y V_Y}{V_M+V_Y}\Rightarrow[Y]=\frac{\dfrac{c_Y V_Y}{V_M+V_Y}}{1+K_{MY}[M]} \tag{7-28}$$

Now substitute the expression for [Y] from Equation 7-28 back into Equation 7-27:

$$[M]\left(1+K_{MY}\frac{\dfrac{c_Y V_Y}{V_M+V_Y}}{1+K_{MY}[M]}\right)=\frac{c_M V_M}{V_M+V_Y} \tag{7-29}$$

and do about fifive lines of algebra to solve for the fraction of titration, ϕ:

Spreadsheet equation for titration of M with Y:

$$\phi=\frac{c_Y V_Y}{c_M V_M}=\frac{1+K_{MY}[M]-\dfrac{[M]+K_{MY}[M]^2}{c_M}}{K_{MY}[M]+\dfrac{[M]+K_{MY}[M]^2}{c_Y}} \tag{7-30}$$

As in acid-base titrations, ϕ is the fraction of the way to the equivalence point. When ϕ=1, V_Y=V_e. When ϕ=1/2, V_Y=1/2 V_e. And so on.

For a titration with EDTA, you can follow the derivation through and fifind that the formation constant, K_{MY}, should be replaced in Equation 7-30 by the conditional formation constant, K'_{MY}, which applies at the fixed pH of the titration.

If you reverse the process and titrate ligand with metal ion, the fraction of the way to the equivalence point is the inverse of the fraction in Equation 7-30:

Spreadsheet equation for titration of Y with M:

$$\phi=\frac{c_M V_M}{c_Y V_Y}=\frac{K_{MY}[M]+\dfrac{[M]+K_{MY}[M]^2}{c_Y}}{1+K_{MY}[M]-\dfrac{[M]+K_{MY}[M]^2}{c_M}} \tag{7-31}$$

Let's calculate the shape of the titration curve for the reaction of 50.0ml of 0.0400mol/L Ca^{2+}(buffered to pH 10.00) with 0.0800mol/L EDTA:

$$Ca^{2+}+Y^{4-}\rightleftharpoons CaY^{2-}$$

$$K'_{CaY}=\delta_{Y^{4-}}K_{CaY}=(0.30)(10^{10.65})=1.34\times10^{10}$$

Because K'_{CaY} is large, it is reasonable to say that the reaction goes to completion with each addition of titrant. We want to make a graph in which is plotted versus milliliters of added EDTA. The equivalence

volume is 25.0ml.

Region 1: Before the equivalence point

Consider the addition of 5.0ml of EDTA. Because the equivalence point requires 25.0ml of EDTA, one-fifth of the Ca^{2+} will be consumed and four-fifths remains.

$$[Ca^{2+}]=\left(\frac{25.0-5.0}{25.0}\right)(0.0400)\left(\frac{50.0}{55.0}\right)$$

$$=0.0291\text{mol/L}$$

$$\Rightarrow pCa^{2+}=-\lg[Ca^{2+}]=1.54$$

In a similar manner, we could calculate pCa^{2+} for any volume of EDTA less than 25.0ml.

Region 2: At the equivalence point

Virtually all the metal is in the form CaY^{2-}. Assuming negligible dissociation, we find the concentration of CaY^{2-} to be equal to the original concentration of Ca^{2+} with a correction for dilution.

$$[CaY^{2-}]=(0.0400)\left(\frac{50.0}{75.0}\right)$$

$$=0.0267\text{mol/L}$$

The concentration of free Ca^{2+} is small and unknown. We can write

	Ca^{2+}	$+Y^{4-}$	$\rightleftharpoons CaY^{2-}$
Initial concentration (mol/L)	0	0	0.0267
Final concentration (mol/L)	x	x	$0.0267-x$

$$\frac{[CaY^{2-}]}{[Ca^{2+}][Y^{4-}]}=K'_{MY}=1.34\times10^{10}$$

$$\frac{0.0267-x}{x^2}=1.34\times10^{10}\Rightarrow x=1.4\times10^{-6}\text{mol/L}$$

$$pCa^{2+}=-\lg x=5.85$$

Region 3: After the equivalence point

In this region, virtually all the metal is in the form CaY^{2-}, and there is excess, unreacted EDTA. The concentrations of CaY^{2-} and excess EDTA are easily calculated. For example, at 26.0ml, there is 1.0ml of excess EDTA.

$$[Y^{4-}]=(0.0800)\left(\frac{1.0}{76.0}\right)=1.05\times10^{-3}\text{mol/L}$$

$$[CaY^{2-}]=(0.0400)\left(\frac{50.0}{76.0}\right)=2.63\times10^{-2}\text{mol/L}$$

The concentration of Ca^{2+} is governed by

$$\frac{[CaY^{2-}]}{[Ca^{2+}][Y^{4-}]}=K'_{CaY}=1.34\times10^{10}$$

$$\frac{[2.63\times10^{-2}]}{[Ca^{2+}](1.05\times10^{-3})}=1.34\times10^{10}$$

$$[Ca^{2+}]=1.9\times10^{-9}\text{mol/L}\Rightarrow pCa^{2+}=8.73$$

The same sort of calculation can be used for any volume past the equivalence point.

Calculated titration curves for Ca^{2+} and Sr^{2+} in Figure 7-6 show a distinct break at the equivalence point, where the slope is greatest. The Ca^{2+} end point is more distinct than the Sr^{2+} end point because the conditional formation constant, K'_{MY}, for CaY^{2-} is greater than that of SrY^{2-}. If the pH is lowered, the conditional formation constant decreases (because $\delta_{Y^{4-}}$ decreases), and the end point becomes less distinct. The pH cannot be raised arbitrarily high because metal hydroxide might precipitate.

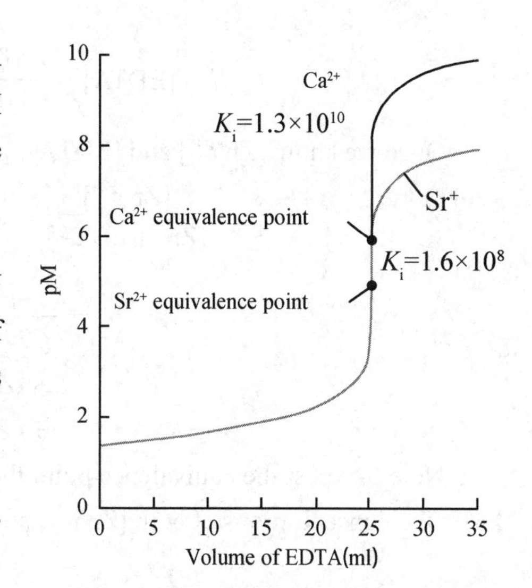

Figure 7-6 The oretical titration curves for the reaction of 50.0ml of 0.0400mol/L metal ion with 0.0800mol/L EDTA at pH 10.00

【Example 7-6】

Consider the titration of 50.0ml of 1.00×10^{-3}mol/L Zn^{2+} with 1.00×10^{-3}mol/L EDTA at pH 10.00 in the presence of 0.10mol/L NH_3. (This is the concentration of NH_3. There is also NH_4^+ in the solution.) The equivalence point is at 50.0ml. Find pZn^{2+} after addition of 20.0, 50.0, and 60.0ml of EDTA.

Solution: In Equation 7-20, we found that $\delta_{Zn^{2+}}=1.8\times10^{-5}$, Table 7-2 tells us that $\delta_{Y^{4-}}=0.30$.

Therefore, the conditional formation constant is

$$K'_{ZnY}=\delta_{Zn^{2+}}\delta_{Y^{4-}}K_{ZnY}=(1.8\times10^{-5})(0.30)(10^{16.5})=1.7\times10^{11}$$

(a) Before the equivalence point—20.0ml: Because the equivalence point is 50.0ml, the fraction of Zn^{2+} remaining is 30.0/50.0. The dilution factor is 50.0/70.0.

Therefore, the concentration of zinc not bound to EDTA is

$$c_{Zn^{2+}}=\left(\frac{30.0}{50.0}\right)(1.00\times10^{-3})\left(\frac{50.0}{70.0}\right)$$
$$=4.3\times10^{-4}\text{mol/L}$$

However, nearly all zinc bound to EDTA is bound to NH_3, The concentration of free Zn^{2+} is

$$[Zn^{2+}]=\delta_{Zn^{2+}}c_{Zn^{2+}}=(1.8\times10^{-5})(4.3\times10^{-4})=7.7\times10^{-9}\text{mol/L}$$
$$\Rightarrow pZn^{2+}=-\lg[Zn^{2+}]=8.11$$

Let's try a reality check. The product $[Zn^{2+}][OH^-]^2$ is $[10^{-8.11}][10^{-4.00}]^2=10^{-16.11}$, which does not exceed the solubility product of $Zn(OH)_2$ ($K_{sp}=10^{-15.52}$).

(b) At the equivalence point—50.0ml: At the equivalence point, the dilution factor is 50.0/100.0, so $[ZnY^{2-}]=(50.0/100.0)(1.00\times10^{-3})=5.00\times10^{-4}$mol/L. We write

$$Zn^{2+}+EDTA\rightleftharpoons ZnY^{2-}$$

	Zn^{2+}	EDTA	ZnY^{2-}
Initial concentration (mol/L)	0	0	5.00×10^{-4}
Final concentration (mol/L)	x	x	$5.00\times10^{-4}-x$

$$K'_{ZnY}=1.7\times10^{11}=\frac{[ZnY^{2-}]}{[c_{Zn^{2+}}][EDTA]}=\frac{5.00\times10^{-4}-x}{x^2}$$
$$\Rightarrow x=c_{Zn^{2+}}=5.4\times10^{-8}\text{mol/L}$$
$$[Zn^{2+}]=\delta_{Zn^{2+}}c_{Zn^{2+}}=(1.8\times10^{-5})(5.4\times10^{-8})=9.7\times10^{-13}\text{mol/L}$$
$$\Rightarrow pZn^{2+}=-\lg[Zn^{2+}]=12.01$$

(c) After the equivalence point—60.0ml: Almost all zinc is in the form with a dilution factor of 50.0/110.0 for zinc, we find

$$[Zn^{2+}]=\left(\frac{50.0}{110.0}\right)(1.00\times10^{-3})=4.5\times10^{-4}\text{mol/L}$$

We also know the concentration of excess EDTA, whose dilution factor is 10.0/110.0:

$$[EDTA]=\left(\frac{10.0}{110.0}\right)(1.00\times10^{-3})=9.1\times10^{-5}\text{mol/L}$$

Once we know $[ZnY^{2-}]$ and [EDTA], we can use the equilibrium constant to find $[Zn^{2+}]$:

$$\frac{[ZnY^{2-}]}{[Zn^{2+}][EDTA]}=\delta_{Y^{4-}}\cdot K_{ZnY}=K'_{ZnY}=(0.30)(10^{16.5})=9.5\times10^{15}$$

$$\frac{[4.5\times10^{-4}]}{[Zn^{2+}][9.1\times10^{-5}]}=9.1\times10^{15}$$

$$\Rightarrow [Zn^{2+}]=5.2\times10^{-16}\text{mol/L}$$

$$\Rightarrow pZn^{2+}=15.28$$

Note that past the equivalence point the problem did not depend on the presence of NH_3, because we knew the concentrations of both $[ZnY^{2-}]$ and [EDTA].

5. Metal ion indicator

We can measure the pM potentiometrically if a suitable electrode is available, for example, an ion-selective electrode (see Chapter 9), but it is simpler if an indicator can be used. Indicators used for complexometric titrations are themselves chelating agents. They are usually dyes of the *o*, *o*'-dihydroxy azo type, known as metal ion indicators (金属离子指示剂).

5.1 The function principle of indicator and the necessary conditions

5.1.1 How indicators work

Metal ion indicators are compounds whose color changes when they bind to a metal ion. Useful indicators must bind metal less strongly than EDTA does. Most indicators for complexation titrations are organic dyes that form stable complexes with metal ions. To function as an indicator for an EDTA titration, the metal-indicator complex must possess a color different from that of the uncomplexed indicator. Furthermore, the formation constant for the metal–indicator complex must be less favorable than that for the metal-EDTA complex.

The indicator is added to the solution of analyte, forming a colored metal-indicator complex. As EDTA is added, it reacts first with the free analyte, and then displaces the analyte from the metal-indicator complex, affecting a change in the solution's color. Eriochrome Black T (铬黑 T, EBT or ET) is a typical indicator. It contains three ionizable protons, so we will represent it by H_3In. This indicator can be used for the titration of Mg^{2+} with EDTA. A small amount of indicator is added to the sample solution, and it forms a red complex with part of the Mg^{2+}; the color of the uncomplexed indicator is blue. As soon as all the free Mg^{2+} is titrated, the EDTA displaces the indicator from the magnesium, causing a change in the color from red to blue:

$$\underset{\text{(red)}}{MgIn^{-}} + \underset{\text{(colorless)}}{H_2Y^{2-}} \rightleftharpoons \underset{\text{(colorless)}}{MgY^{2-}} + \underset{\text{(blue)}}{HIn^{2-}} + H^{+}$$

This will occur over a range of pMg values, and the change will be sharper if the indicator is kept as dilute as possible but is still sufficient to give a good color change.

5.1.2 Essential conditions for metal indicator

Most metallochromic indicators also are weak acids or bases. The conditional formation constant for the metal–indicator complex, therefore, depends on the solution's pH. This provides some control over the indicator's titration error (滴定误差).

The apparent strength of a metal-indicator complex can be adjusted by controlling the pH at which

the titration is carried out. Unfortunately, because they also are acid-base indicators, the color of the uncomplexed indicator changes with pH. For example, calmagite, which we may represent as H_3In, undergoes a change in color from the red of H_2In^- to the blue of HIn^{2-} at a pH of approximately 8.1, and from the blue of HIn^{2-} to the red-orange of In^{3-} at a pH of approximately 12.4. Since the color of calmagite's metal-indicator complexes are red, it is only useful as a metal lochromic indicator in the pH range of 9–11, at which almost all the indicator is present as HIn^{2-}.

The accuracy of the end point depends on the strength of the metal-indicator complex relative to that of the metal-EDTA complex. The metal-indicator complex must be less stable than the metal-EDTA complex, or else the EDTA will not displace it from the metal. On the other hand, it must not be too weak, or the EDTA will start replacing it at the beginning of the titration, and a diffuse end point will result. In general, the *K* for the metal–indicator complex should be 10 to 100 times less than that for the metal-titrant complex.

5.2 Common metal indicators

A partial list of Metal ion indicators, and the metal ions for which they are useful, is given in Table 7-3, For example, Xylenol orange, XO (二甲酚橙); Calcon-carboxylic, NN (钙指示剂) .

Table 7-3 Common metal ion indicators

Name	pK_a	Color of free indicator		Color of metal ion complex
Calmagine	pK_2 = 8.1 pK_3 = 12.4	H_2In^- HIn^{2-} In^{3-}	red blue orange	Wine red
Eriochrome black T	pK_2 = 6.3 pK_3 = 11.6	H_2In^- HIn^{2-} In^{3-}	red blue orange	Wine red
Murexide	pK_2 = 9.2 pK_3 = 10.9	H_4In^- H_3In^{2-} H_2In^{3-}	red-violet violet blue	Yellow (with Co^{2+},Ni^{2+}, Cu^{2+}); red with Ca^{2+}
Xylenol orange	pK_2 = 2.32 pK_3 = 2.85 pK_4 = 6.70 pK_5 = 10.47 pK_6 = 12.23	H_5In^- H_4In^{2-} H_3In^{3-} H_2In^{4-} HIn^{5-} In^{6-}	yellow yellow yellow violet violet violet	Red
Pyrocatechol violet	pK_1 = 0.2 pK_2 = 7.8 pK_3 = 9.8 pK_4 = 11.7	H_4In H_3In^- H_2In^{2-} HIn^{3-}	red-violet violet blue red-purple	Blue

Most metal ion indicators are also acid-base indicators (酸碱指示剂), Because the color of free indicator is pH dependent, most indicators can only be used in certain pH ranges. Figure 7-7 shows pH ranges in which many metals can be titrated and indicators that are useful in different ranges.

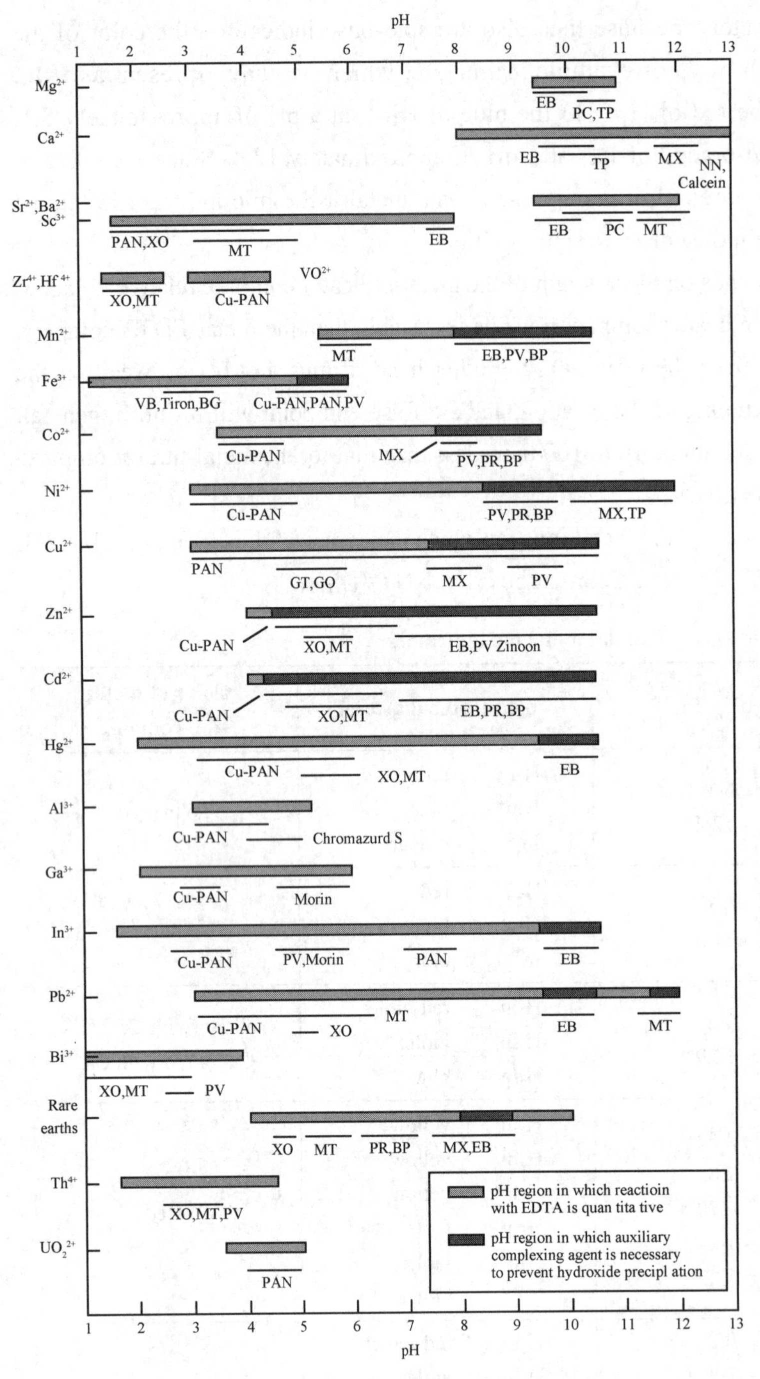

Figure 7-7 Guide to EDTA titrations of common metals

NOTE: The Light color shows pH region in which reaction with EDTA is quantitative. Dark color shows pH region in which auxiliary complexing agent is required to prevent metal from precipitating. Calmagite is more stable than Eriochrome black T (EB) and can be substituted for EB.

Abbreviations for indicators:

BG. Bindschedler's green leuco base
BP. Bromopyrogallol red
EB. Eriochrome black T
GC. Glycinecresol red
GT. Glycinethymol blue
MT. Methylthymol blue
MX. Murexide
NN. Patton & Reeder's dye
PAN. Pyridylazonaphthol
Cu-PAN. PAN plus Cu-EDTA
PC. *o*-Cresolphthalein complexone
PR. Pyrogallol red
PV. Pyrocatechol violet
TP. Thymolphthalein complexone
VB. Variamine blue B base
XO. Xylenol orange

For an indicator to be useful in the titration of a metal with EDTA. the indicator must give its metal ion to the EDTA. If a metal does not freely dissociate from an indicator, the metal is said to block the indicator. EBT is blocked by Cu^{2+}, Ni^{3+}, Co^{2+}, Cr^{3+}, Fe^{3+} and Al^{3+}. It cannot be used for the direct titration of any of these metals, but can be used for a back titration. For example, excess standard EDTA can be added to Cu^{2+}. Then indicator is added and the excess EDTA is back-tatrated with Mg^{2+}.

Section 2 Improving the Selectivity of Complexometric Titration

1. Conditions to eliminate the effects of interfering ions

In analytical practice, the samples to be analyzed are frequently complex and more than one metal ion may be present in the solution. Suppose that in a solution containing two metal ions. M and N, both the ions can react with EDTA forming complexes. If their concentrations are equal and $K'_{MY} > K'_{NY}$, and then M will be titrated with EDTA first. As the difference between K'_{MY} and K'_{NY} is large enough. M will quantitatively react with EDTA before N begins to be titrated. That is, M can be accurately titrated with EDTA in the presence of N. The feasibility of successive titration of the two metal ions depends not only on the difference between K'_{MY} and K'_{NY}, but also on the concentration ratio, c_M/c_N. It has been proved that if M can be accurately titrated (准确滴定) in the presence of N, certain requirements must be satisfied, such as

$$\lg c_M K'_{MY} > 6$$

$$\text{and} \quad \lg\frac{c_M K_{MY}}{c_N K_{NY}} = \Delta \lg K + \lg\left(\frac{c_M}{c_N}\right) \geqslant 5 \tag{7-32}$$

When $c_M = c_N$, Equation 7-32 can be simplified as

$$\Delta \lg K \geqslant 5 \tag{7-33}$$

For example, in a solution containing 0.01mol/L Bi^{3+}, and 0.01mol/L Pb^{2+}, consider whether the two metal ions can be successively titrated. Since $\lg K_{BiY}$=27.94 and $\lg K_{PbY}$=18.04. then $\Delta \lg K$=9.90. This indicates that Bi^{3+} can be selectively titrated (选择性滴定) in a suitable pH range and Pb^{2+} does not interfere with the determination. The minimum pH value (最小pH值) for titration of Bi^{3+} is estimated to be 0.7. That is, Bi^{3+} should be titrated at pH>0.7. On the other hand, the titration should be performed at pH<2 (usually at pH=1) in order to prevent Bi^{3+} from hydrolysis. In this case, $\lg K'_{PbY}$=0. 03 and Pb^{2+} does not react with EDTA. After Bi^{3+} has been completely titrated, the acidity of the solution is then adjusted to pH 6 ($\lg K'_{PbY}$=13.4), and then Pb^{2+} can be successively titrated with EDTA.

2. Measures to improve the selectivity of coordination titration-Masking interfering ions

2.1 Masking

A masking agent (掩蔽剂) is a reagent that protects some component of the analyte from reaction with EDTA. For example, Al^{3+} in a mixture of Mg^{2+} and Al^{3+} can be measured by first masking the Al^{3+} with F^-, thereby leaving only the Mg^{2+} to react with EDTA.

Cyanide masks Cd^{2+}, Zn^{2+}, Hg^{2+}, Co^{2+}, Cu^+, Ag^+, Ni^{2+}, Pd^{2+}, Pt^{2+}, Fe^{2+}, and Fe^{3+}, but not Mg^{2+}, Ca^{2+}, Mn^{2+}, or Pb^{2+}. When cyanide is added to a solution containing Cd^{2+} and Pb^{2+}, only Pb^{2+} reacts with EDTA. (Caution: Cyanide forms toxic gaseous HCN below pH 11. Cyanide solutions should be strongly basic and only handled in a hood.) Fluoride masks Al^{3+}, Fe^{3+}, Ti^{4+}, and Be^{2+}. (Caution: HF formed by F^- in

acidic solution is extremely hazardous and should not contact skin and eyes. It may not be immediately painful, but the affected area should be flooded with water and then treated with calcium gluconate gel that you have on hand before the accident. First aid providers must wear rubber gloves to protect themselves.) Triethanolamine masks Al^{3+}, Fe^{3+}, and Mn^{2+}; 2,3-dimercapto-1-propanol masks Bi^{3+}, Cd^{2+}, Cu^{2+}, Hg^{2+}, and Pb^{2+}.

2.2 Demasking

releases metal ion from a masking agent. Cyanide complexes can be demasked with formaldehyde:

$$M(CN)_m^{n-m} + mH_2CO + mH^+ \rightleftharpoons mH_2C(OH)(CN) + M^{n+}$$

Thiourea masks Cu^{2+} by reducing it to Cu^+ and complexing the Cu^+. Copper can be liberated from thiourea by oxidation with H_2O_2. Selectivity afforded by masking, demasking (解掩蔽作用), and pH control allows individual components of complex mixtures of metal ions to be analyzed by EDTA titration.

Section 3 Complexometric Titration Method and Its Application

1. Different ways of titrimetric analysis

Because so many elements can be analyzed with EDTA, there is extensive literature dealing with many variations of the basic procedure.

1.1 Direct titration

In a direct titration (直接滴定法), analyte is titrated with standard EDTA. Analyte is buffered to a pH at which the conditional formation constant for the metal-EDTA complex is large and the color of the free indicator is distinctly different from that of the metal-indicator complex.

An auxiliary complexing agent (辅助配位剂), such as ammonia, tartrate, citrate, or triethanolamine (三乙醇胺), may be employed to prevent the metal ion from precipitating in the absence of EDTA.

For example, the direct titration of Pb^{2+} is carried out in ammonia buffer at pH 10 in the presence of tartrate, which complexes the metal ion and does not allow $Pb(OH)_2$ to precipitate. The lead-tartrate complex must be less stable than the lead-EDTA complex, or the titration would not be feasible.

1.2 Back titration

In a back titration (返滴定法), a known excess of EDTA is added to the analyte. Excess EDTA is then titrated with a standard solution of a second metal ion. A back titration is necessary if analyte precipitates in the absence of EDTA, if it reacts too slowly with EDTA, or if it blocks the indicator. The metal ion for the back titration must not displace analyte from EDTA.

【Exemple 7-7】

Ni^{2+} can be analyzed by a back titration using standard Zn^{2+} at pH 5.5 with xylenol orange indicator. A solution containing 25.00ml of Ni^{2+} in dilute HCl is treated with 25.00ml of 0.05283mol/L Na_2EDTA. The solution is neutralized with NaOH, and the pH is adjusted to 5.5 with acetate buffer. The solution turns yellow when a few drops of indicator are added. Titration with 0.02299mol/L Zn^{2+} requires 17.61ml to reach the red end point. What is the molarity of Ni^{2+} in the unknown?

Solution: The unknown was treated with 25.00ml of 0.05283mol/L EDTA, which contains (25.00ml) (0.05283mol/L) = 1.3208mmol of EDTA. Back titration required (17.61ml) × (0.02299mol/L)= 0.4049mmol of Zn^{2+}. Because 1mol of EDTA reacts with 1mol of any metal ion, there must have been

$$1.3208\text{mmol EDTA} - 0.4049\text{mmol } Zn^{2+} = 0.9159\text{mmol } Ni^{2+}$$

The concentration of Ni^{2+} is 0.9159mmol/25.00ml=0.03664mol/L.

Back titration prevents precipitation of analyte. For example, $Al(OH)_3$ precipitates at pH 7 in the absence of EDTA. An acidic solution of Al^{3+} can be treated with excess EDTA, adjusted to pH 7–8 with sodium acetate, and boiled to ensure complete formation of stable, soluble $Al(EDTA)^-$. The solution is then cooled; Calmagite indicator is added; and back titration with standard Zn^{2+} is performed.

1.3 displacement titration

Hg^{2+} does not have a satisfactory indicator, but a displacement titration (置换滴定法) is feasible. Hg^{2+} is treated with excess $Mg(EDTA)^{2-}$ to displace Mg^{2+}, which is titrated with standard EDTA.

$$M^{n+} + MgY^{2-} \rightleftharpoons MY^{n-4} + Mg^{2+} \quad (7\text{-}34)$$

The conditional formation constant for HgY^{2-} must be greater than K'_{MY} for MgY^{2-}, or else Mg^{2-} will not be displaced from MgY^{2-}.

There is no suitable indicator for Ag^+. However, Ag^+ will displace Ni^{2+} from tetracyanonickelate (II) ion:

$$2Ag^+ + Ni(CN)_4^{2-} \rightleftharpoons 2Ag(CN)_2^- + Ni^{2+}$$

The liberated Ni^{2+} can then be titrated with EDTA to find out how much Ag^+ was added.

1.4 Indirect titration

Anions that precipitate with certain metal ions can be analyzed with EDTA by indirect titration (间接滴定法). For example, sulfate can be analyzed by precipitation with excess at pH 1. The $BaSO_4(s)$ is washed and then boiled with excess EDTA at pH 10 to bring Ba^{2+} back into solution as $Ba(EDTA)^{2-}$. Excess EDTA is back-titrated with Mg^{2+}.

Alternatively, an anion can be precipitated with excess standard metal ion. The precipitate is filtered and washed, and excess metal in the filtrate is titrated with EDTA. Anions such as CO_3^{2-}, CrO_4^{2-}, S^{2-}, and SO_4^{2-} can be determined by indirect titration with EDTA.

2. Application examples

Determination of Water Hardness

Historically, water "hardness" was defined in terms of the capacity of cations in the water to replace the sodium or potassium ions in soaps and form sparingly soluble products that cause "scum" in the sink or bathtub. Most multiply charged cations share this undesirable property. In natural waters, however, the concentrations of calcium and magnesium ions generally far exceed those of any other metal ion. Consequently, hardness is now expressed in terms of the concentration of calcium carbonate that is equivalent to the total concentration of all the multivalent cations in the sample.

The determination of hardness is a useful analytical test that provides a measure of the quality of water for household and industrial uses. The test is important to industry because hard water, on being heated, precipitates calcium carbonate, which clogs boilers and pipes.

Water hardness is usually determined by an EDTA titration after the sample has been buffered to pH 10. Magnesium, which forms the least stable EDTA complex of all of the common multivalent cations in typical water samples, is not titrated until enough reagent has been added to complex all of the other cations in the sample. Therefore, a magnesium-ion indicator, such as Calmagite or Eriochrome Black T, can serve as indicator in water-hardness titrations. Often, a small concentration of the magnesium-EDTA chelate is incorporated in the buffer or in the titrant to ensure the presence of sufficient magnesium ions for satisfactory indicator action.

重点小结

配位滴定法又称络合滴定法，是以配位反应为基础的滴定分析法。此法主要用于金属离子的测定，多数金属离子在溶液中以配位离子的形式存在。EDTA是广泛用于滴定金属离子的氨羧配位剂，形成的配位化合物具有稳定性高、配位比简单、反应速度快、颜色多为无色等优点。表征配位化合物稳定性的参数为稳定常数及条件稳定常数，两者的区别是条件稳定常数考虑副反应的影响，所以，条件稳定常数更能反映实际情况。事实上，配位反应涉及化学平衡很复杂，除了被测金属离子与滴定剂之间的主反应外，往往有影响主反应的一些其他副反应存在，采用副反应系数定量地表示副反应进行的程度。

在配位滴定中，通常采用金属离子指示剂判定终点的到达。金属离子指示剂是一种有机染料，它可以与被滴定金属离子形成一种与染料本身颜色不同的配合物。金属指示剂应具备以下条件：① 生成配合物颜色与知识及颜色有明显区别；②金属配合物的稳定性比金属 -EDTA 配合物的稳定性低。

在配位滴定中，由于酸度对金属离子、EDTA及指示剂都可能产生影响，所以应考虑滴定条件的选择和控制。具体条件分两种情况：① 对于单一离子的滴定条件应考虑最高酸度（$\lg\alpha_{YM}=\lg K_{MY}-8$）及最低酸度（$[OH^-]\geq\sqrt[n]{K_{sp}/c_M}$）的酸度条件；②对于混合离子的选择性滴定可以通过控制酸度（$\Delta\lg K\geq5$）或利用掩蔽法进行控制。常用的掩蔽方法有配位掩蔽法、沉淀掩蔽法和氧化还原掩蔽法。

配位滴定方式有直接滴定法、返滴定法、置换滴定法和间接滴定法等类型。由于这些方法的应用，配位滴定能够直接或间接测定周期表中的大多数元素。

题库

目标检测

1. Define

(a) ligand

(b) chelate

(c) tetradentate chelating agent

(d) adsorption indicator

(e) conditional formation constant

(f) EDTA displacement titration

(g) water hardness

2. Describe three general methods for performing EDTA titrations. What are the advantages of each?

3. Write chemical equations and equilibrium-constant expressions for the stepwise formation of

(a) $Ag(S_2O_3)_2^{3-}$

(b) $Ni(CN)_4^{2-}$

(c) $Cd(SCN)_3^-$

4. Explain how stepwise and overall formation constants are related.

5. Write chemical formulas for the following complex ions:

(a) hexamminezinc(Ⅱ)

(b) dichloroargentate

(c) disulfatocuprate(Ⅱ)

(d) trioxalotoferrate(Ⅲ)

(e) hexacyanoferrate(Ⅱ)

6. An EDTA solution was prepared by dissolving 3.426g of purified and dried $Na_2H_2Y_2 \cdot 2H_2O$ in sufficient water to give 1.000L. Calculate the molar concentration, given that the solute contained 0.3% excess moisture.

7. A solution contains 1.569mg of $CoSO_4$ (155.0g/mol) per milliliter. Calculate

(a) the volume of 0.007840mol/L EDTA needed to titrate a 25.00ml aliquot of this solution.

(b) the volume of 0.009275mol/L Zn^{2+} needed to titrate the excess reagent after addition of 50.00ml of 0.007840mol/L EDTA to a 25.00ml aliquot of this solution.

(c) the volume of 0.007840mol/L EDTA needed to titrate the Zn^{2+} displaced by Co^{2+} following addition of an unmeasured excess of ZnY^{2-} to a 25.0ml aliquot of the $CoSO_4$ solution. The reaction is $Co^{2+} + ZnY^{2-} \rightarrow CoY^{2-} + Zn^{2+}$

8. Calculate the volume of 0.0500mol/L EDTA needed to titrate

(a) 29.13ml of 0.0598mol/L $Mg(NO_3)_2$.

(b) the Ca in 0.1598g of $CaCO_3$.

(c) the Ca in a 0.4861g mineral specimen that is 81.4% brushite, $CaHPO_4 \cdot 2H_2O$ (172.09g/mol).

(d) the Mg in a 0.1795g sample of the mineral hydromagnesite, $Mg_4(OH)_2(CO_3)_3 \cdot 3H_2O$ (365.3g/mol).

(e) the Ca and Mg in a 0.1612g sample that is 92.5% dolomite, $CaCO_3 \cdot MgCO_3$ (184.4g/mol).

9. The Zn in a 0.7457g sample of foot powder was titrated with 22.57ml of 0.01639mol/L EDTA. Calculate the percent Zn in this sample.

10. The Tl in a 9.57g sample of rodenticide was oxidized to the trivalent state and treated with an unmeasured excess of Mg/EDTA solution. The reaction is

$$Tl^{3+} + MgY^{2-} \rightarrow TlY^- + Mg^{2+}$$

Titration of the liberated Mg^{2+} required 12.77ml of 0.03610mol/L EDTA. Calculate the percent Tl_2SO_4 (504.8g/mol) in the sample.

11. An EDTA solution was prepared by dissolving approximately 4g of the disodium salt in approximately 1L of water. An average of 42.35ml of this solution was required to titrate 50.00ml aliquots of a standard that contained 0.7682g of $MgCO_3$ per liter. Titration of a 25.00ml sample of mineral water at pH 10 required 18.81ml of the EDTA solution. A 50.00ml aliquot of the mineral water was rendered strongly alkaline to precipitate the magnesium at $Mg(OH)_2$. Titration with a calcium-specific indicator required 31.54ml of the EDTA solution. Calculate

(a) the molar concentration of the EDTA solution.

(b) the concentration of $CaCO_3$ in the mineral water in ppm.

(c) the concentration of $MgCO_3$ in the mineral water in ppm.

12. The formation constant of the silver–ethylenediamine complex, $Ag(NH_2CH_2CH_2NH_2)^+$, is 5.0×10^4. Calculate the concentration of Ag^+ in equilibrium with a 0.mol/L solution of the complex. (Assume no higher order complexes.)

13. What would be the concentration of Ag^+ in Problem 12 if the solution contained also 0.10mol/L ethylenediamine, $NH_2CH_2CH_2NH_2$?

14. Silver ion forms stepwise complexes with thiosulfate ion, $S_2O_3^{2-}$, with $K_1 = 6.6 \times 10^8$ and $K_2 = 4.4 \times 10^4$. Calculate the equilibrium concentrations of all silver species for 0.0100mol/L $AgNO_3$ in 1.00mol/L $Na_2S_2O_3$. Neglect diverse ion effects.

15. A 50.00ml aliquot of a solution containing iron(Ⅱ) and iron(Ⅲ) required 10.98ml of 0.01500mol/L EDTA when titrated at pH 2.0 and 23.70ml when titrated at pH 6.0. Express the concentration of each solute (mg/L).

16. A 24-hr urine specimen was diluted to 2.000L. After the solution was buffered to pH 10, a 10.00ml aliquot was titrated with 23.57ml of 0.004590mol/L EDTA. The calcium in a second 10.00ml aliquot was isolated as $CaC_2O_4(s)$, redissolved in acid, and titrated with 10.53ml of the EDTA solution. Assuming that 15 to 300mg of magnesium and 50 to 400mg of calcium per day are normal, did this specimen fall within these ranges?

17. A 1.509g sample of a Pb/Cd alloy was dissolved in acid and diluted to exactly 250.0ml in a volumetric flask. A 50.00ml aliquot of the diluted solution was brought to a pH of 10.0 with a NH_4^+/NH_3 buffer; the subsequent titration involved both cations and required 28.89ml of 0.06950mol/L EDTA. A second 50.00ml aliquot was brought to a pH of 10.0 with an HCN/NaCN buffer, which also served to mask the Cd^{2+}; 11.56ml of the EDTA solution were needed to titrate the Pb^{2+}. Calculate the percent Pb and Cd in the sample.

18. A 0.6004g sample of Ni/Cu condenser tubing was dissolved in acid and diluted to 100.0ml in a volumetric flask. Titration of both cations in a 25.00ml aliquot of this solution required 45.81ml of 0.05285mol/L EDTA. Mercaptoacetic acid and NH_3 were then introduced; production of the Cu complex with the former resulted in the release of an equivalent amount of EDTA, which required a 22.85ml titration with 0.07238mol/L Mg^{2+}. Calculate the percent Cu and Ni in the alloy.

19. The formation constant for the lead-EDTA chelate (PbY^{2-}) is 1.10×10^{18}. Calculate the conditional formation constant (a) at pH 3 and (b) at pH 10.

20. Using the conditional constants calculated in Problem 19 calculate the pPb ($-\lg[Pb^{2+}]$) for 50.0ml of a solution of 0.0250mol/L Pb^{2+} (a) at pH 3 and (b) at pH 10 after the addition of (1) 0ml, (2) 50ml, (3) 125ml, and (4) 200ml of 0.0100mol/L EDTA.

21. The conditional formation constant for the calcium-EDTA chelate was calculated for pH 10 to be 1.8×10^{10}. Calculate the conditional formation constant at pH 3. Compare this with that calculated for lead at pH 3 in Problem 19. Could lead be titrated with EDTA at pH 3 in the presence of calcium?

22. Calculate the conditional formation constant for the nickel-EDTA chelate in an ammoniacal pH 10 buffer containing $[NH_3]$=0.10mol/L.

23. Calamine, which is used for relief of skin irritations, is a mixture of zinc and iron oxides. A 1.056g sample of dried calamine was dissolved in acid and diluted to 250.0ml. Potassium fluoride was added to a 10.00ml aliquot of the diluted solution to mask the iron; after suitable adjustment of the pH, Zn^{2+} consumed 38.37ml of 0.01133mol/L EDTA. A second 50.00ml aliquot was suitably buffered and titrated with 2.30ml of 0.002647mol/L ZnY^{2-} solution:

$$Fe^{3+} + ZnY^{2-} \rightarrow FeY^{-} + Zn^{2+}$$

Calculate the percentages of ZnO and Fe_2O_3 in the sample.

24. Chromel is an alloy composed of nickel, iron, and chromium. A 0.6553g sample was dissolved and diluted to 250.0ml. When a 50.00ml aliquot of 0.05173mol/L EDTA was mixed with an equal volume of the diluted sample, all three ions were chelated, and a 5.34ml back-titration with 0.06139mol/L copper(II) was required. The chromium in a second 50.0ml aliquot was masked through the addition of hexamethylenetetramine; titration of the Fe and Ni required 36.98ml of 0.05173mol/L EDTA. Iron and chromium were masked with pyrophosphate in a third 50.0ml aliquot, and the nickel was titrated with 24.53ml of the EDTA solution. Calculate the percentages of nickel, chromium, and iron in the alloy.

25. Calculate conditional constants for the formation of the EDTA complex of Fe^{2+} at a pH of (a) 6.0, (b) 8.0, and (c) 10.0.

26. State (in words) what means. Calculate for EDTA at (a) pH 3.50 and (b) pH 10.50.

27. (a) Find the conditional formation constant for at pH 9.00.

(b) Find the concentration of free in 0.050mol/L at pH 9.00.

28. Consider the titration of 25.0ml of 0.0200mol/L with 0.0100mol/L EDTA in a solution buffered to pH 8.00. Calculate at the following volumes of added EDTA and sketch the titration curve:

(a) 0ml (b) 20.0ml (c) 40.0ml (d) 49.0ml (e) 49.9ml

(f) 50.0ml (g) 50.1ml (h) 55.0ml (i) 60.0ml

29. Using the same volumes as in Problem 28, calculate for the titration of 25.00ml of 0.02000mol/L EDTA with 0.01000mol/L at pH 10.00.

30. Calculate the molarity of in a solution prepared by mixing 10.00ml of 0.0100mol/L 9.90ml of 0.0100mol/L EDTA, and 10.0ml of buffer with a pH of 4.00.

31. Calculate at each of the following points in the titration of 50.00mL of 0.00100mol/L with 0.00100mol/L EDTA at pH 11.00 in a solution whose concentration is somehow fixed at 0.100mol/L:

(a) 0ml

(b) 1.00ml

(c) 45.00ml

(d) 50.00ml

(e) 55.00ml

32. List four methods for detecting the end point of an EDTA titration.

33. Calcium ion was titrated with EDTA at pH 11, using Calmagite as indicator. Which is the principal species of Calmagite at pH 11? What color was observed before the equivalence point? After the equivalence point?

34. Pyrocatechol violet (see Table 7-3) is to be used as a metal ion indicator in an EDTA titration. The procedure is as follows:

(1) Add a known excess of EDTA to the unknown metal ion.

(2) Adjust the pH with a suitable buffer.

(3) Back-titrate the excess chelate with standard Al^{3+}.

From the following available buffers, select the best buffer, and then state what color change will be observed at the end point. Explain your answer.

(a) pH 6–7 (b) pH 7–8 (c) pH 8–9 (d) pH 9–10

35. Titration of Ca^{2+} and Mg^{2+} in a 50.00ml sample of hard water required 23.65ml of 0.01205mol/L

EDTA. A second 50.00ml aliquot was made strongly basic with NaOH to precipitate Mg^{2+} as $Mg(OH)_2(s)$. The supernatant liquid was titrated with 14.53ml of the EDTA solution. Calculate

(a) The total hardness of the water sample, expressed as mg/L $CaCO_3$.

(b) The concentration of $CaCO_3$ in the sample in mg/L.

(c) The concentration of $MgCO_3$ in the sample in mg/L.

36. Calcium ion forms a weak 1 : 1 complex with nitrate ion with a formation constant of 2.0. What would be the equilibrium concentrations of Ca^{2+} and $Ca(NO_3)^+$ in a solution prepared by adding 10ml each of 0.010mol/L $CaCl_2$ and 2.0 mol/L $NaNO_3$? Neglect diverse ion effects.

Chapter 8 Redox Titrations

学习目标

知识要求

1. **掌握** 氧化还原滴定法的原理及结果计算；碘量法的特点及应用。
2. **熟悉** 高锰酸钾法、重铬酸钾法的特点及应用。
3. **了解** 氧化还原反应的特点。

能力要求

1. 能利用条件电位和条件平衡常数进行氧化还原反应进行程度的判断。
2. 掌握影响氧化还原反应速率的因素，从而在实际应用中采用相应措施以提高氧化还原滴定反应的速率。
3. 理解氧化还原滴定法的基本原理，并结合实际能够选择合适的氧化还原滴定分析方法对无机或有机药物进行含量测定。

A redox titration (氧化还原滴定法) is based on an oxidation-reduction reaction between analyte and titrant. Many common analytes in chemistry, biology, environmental and materials science can be measured by redox titrations. For example, the ascorbic acid content of vitamin C tablets, or the content of sulfur dioxide in wines can be measured by titration with iodine. The Karl Fisher titration of water in samples involves iodine-based titrations. The saturation in a fatty acid is determined as the iodine or bromine number. The iron content of an ore can be determined by titration of iron(II) with potassium permanganate or potassium dichromate.

Section 1 Redox Reactions

PPT

A redox reaction occurs between a reducing and an oxidizing agent. An oxidization is defined as a loss of electrons to give a higher oxidation state (more positive), and reduction is defined as a gain of electrons to give a lower oxidation state (more negative). Electrons are transferred from the reducing agent to the oxidizing agent.

1. Formal potential and the influencing factors

1.1 Formal potential

Considering the following redox reaction: $Zn+Cu^{2+} \rightleftharpoons Zn^{2+}+Cu$, it can be arranged as a cell described as:

$$Zn \mid ZnSO_4(x\text{mol/L}) \parallel CuSO_4(y\text{mol/L}) \mid Cu$$

The overall reaction can be broken down into two half-reactions.

At anode (阳极), Zn is oxidized: $Zn \rightleftharpoons Zn^{2+} + 2e$,

and at the cathode (阴极), Cu^{2+} is reduced, $Cu^{2+} + 2e \rightleftharpoons Cu$

The half reaction can be described as: $Ox+ne \rightleftharpoons Red$

Where Ox is the oxidized form (氧化态), and Red is the reduced form (还原态).

The potential of the electrode (电极电位) is obtained by the Nernst equation (能斯特方程):

$$E_{Ox/Red} = E^{\ominus}_{Ox/Red} + \frac{RT}{nF}\ln\frac{a_{Ox}}{a_{Red}} \tag{8-1}$$

Where $E^{\ominus}_{Ox/Red}$ is the standard potential of the electrode (标准电极电位), n is number of electrons involved in the half reaction, R is the gas constant [8.314J/(K·mol)] (气体常数), T is the absolute temperature (绝对温度), F is the Faraday constant (96485C/mol) (法拉第常数). At 25℃, the equation can be described as:

$$E_{Ox/Red} = E^{\ominus}_{Ox/Red} + \frac{0.0592}{n}\ln\frac{a_{Ox}}{a_{Red}} \tag{8-2}$$

The potential of the electrode is dependent on the activities (活度) of the oxidized and reduced forms (and all other species), which are at unit activities, the standard potential $E^{\ominus}$ is obtained. However, the potential of a half reaction may depend on the conditions of the solution. For example, $E^{\ominus}$ for $Ce^{4+} +e \rightleftharpoons Ce^{3+}$ is 1.61 V. The potential $E^{\ominus}$ can be changed by changing the acid for acidifying the solution. The anions of the different acids differ in their ability to form complexes with one form of the cerium (Ce^{4+}) relative to the other (Ce^{3+}), and therefore the activity ratio of a_{Ox}/a_{Red} is affected. If the forms of the complexes are known, a new half reaction involving the acid anion can be written, and $E^{\ominus}$ for this new reaction can be determined by keeping the acid and all other species at unit activity. But, the compositions of all the complexes are frequently unknown. Therefore, the formal potential (or conditional potential) is defined and designed as $E^{\ominus\prime}$. We consider the potential of Fe^{3+}/Fe^{2+} couple in the presence of hydrochloride acid (HCl) as an example.

For the half reaction, $Fe^{3+} + e \rightleftharpoons Fe^{2+}$

$$E = E^{\ominus} + 0.0592\lg\frac{a_{Fe^{3+}}}{a_{Fe^{2+}}}$$

$$= E^{\ominus} + 0.0592\lg\frac{\gamma_{Fe^{3+}}[Fe^{3+}]}{\gamma_{Fe^{2+}}[Fe^{2+}]}$$

In HCl solution, Fe^{3+} ion complexes with Cl^-:

$$Fe^{3+} + Cl^- \rightleftharpoons [FeCl]^{2+}$$

$$[FeCl]^{2+} + Cl^- \rightleftharpoons [FeCl_2]^+$$

$$\vdots$$

The total concentration of Fe(Ⅲ) in the solution can be expressed as

$$c_{Fe^{3+}} = [Fe^{3+}] + [FeCl^{2+}] + [FeCl_2^+] + L$$

The side reaction coefficient of Fe^{3+} designed as $\alpha_{Fe^{3+}}$ (副反应系数 , see the chapter of complexometric titration),

$$\alpha_{Fe^{3+}} = \frac{c_{Fe^{3+}}}{[Fe^{3+}]}, \quad [Fe^{3+}] = \frac{c_{Fe^{3+}}}{\alpha_{Fe^{3+}}}$$

Similarly, for Fe^{2+},

$$\alpha_{Fe^{2+}} = \frac{c_{Fe^{2+}}}{[Fe^{2+}]}, \quad [Fe^{2+}] = \frac{c_{Fe^{2+}}}{\alpha_{Fe^{2+}}}$$

Then, the potential for Fe^{3+}/Fe^{2+} couple can be written as:

$$E = E^{\ominus} + 0.0592 \lg \frac{\gamma_{Fe^{3+}}\alpha_{Fe^{2+}}c_{Fe^{3+}}}{\gamma_{Fe^{2+}}\alpha_{Fe^{3+}}c_{Fe^{2+}}}$$

$$= E^{\ominus} + 0.0592 \lg \frac{\gamma_{Fe^{3+}}\alpha_{Fe^{2+}}}{\gamma_{Fe^{2+}}\alpha_{Fe^{3+}}} + 0.0592 \lg \frac{c_{Fe^{3+}}}{c_{Fe^{2+}}} \qquad (8\text{-}3)$$

When $c_{Fe^{3+}} = c_{Fe^{2+}} = 1mol/L$, substituted in Equation 8-3, we can obtain:

$$E = E^{\ominus} + 0.0592\lg \frac{\gamma_{Fe^{3+}}\alpha_{Fe^{2+}}}{\gamma_{Fe^{2+}}\alpha_{Fe^{3+}}}$$

Under specified conditions, γ (活度系数) and α (副反应系数) is definite and E remains constant, which is then defined as formal potential $E^{\ominus\prime}$ described as:

$$E^{\ominus\prime} = E^{\ominus} + 0.0592\lg \frac{\gamma_{Fe^{3+}}\alpha_{Fe^{2}}}{\gamma_{Fe^{2+}}\alpha_{Fe^{3+}}} \qquad (8\text{-}4)$$

It is obvious that, the formal potential of a redox couple is the potential measured with the total concentrations of the oxidized and reduced forms being both 1mol/L and with the solution conditions being specified. The Nernst equation can be written as usual, using $E^{\ominus\prime}$ instead of $E^{\ominus}$, then Equation 8-3 can be written as:

$$E = E^{\ominus\prime} + 0.0592 \lg \frac{c_{Fe^{3+}}}{c_{Fe^{2+}}} \qquad (8\text{-}5)$$

Accordingly, we can easily get the expression of actual potential for the half reaction $Ox+ne \rightleftharpoons Red$.

$$E = E^{\ominus\prime} + \frac{0.0592}{n} \lg \frac{c_{Ox}}{c_{Red}}$$

In a sense, if the formal potential is available, it is convenient and reasonable to use $E^{\ominus\prime}$ in the equilibrium calculations for redox reactions.

Ionic strength (离子强度) of the solution can affect the potential $E^{\ominus\prime}$, because the activity coefficient γ is much less than 1 for the ions in a solution with high ionic strength. It is difficult to calculate activity coefficient γ. However, the effect of ionic strength of solution on the potential is much less than that of the side reaction, so generally it can be neglected and γ can be approximately equal to 1 for calculations involving Nernst equation.

1.2 The influencing factors

1.2.1 Dependence of formal potential on pH

Many redox reactions involve protons, and their potentials are influenced greatly by pH. The potential of the redox couples can be changed by changing the pH of the solution. Take the As (V)/ As (Ⅲ) as an example:

$$H_3AsO_4 + 2H^+ + 2e \rightleftharpoons H_3AsO_3 + H_2O$$

$$E = E^{\ominus} + \frac{0.0592}{2}\lg\frac{[H_3AsO_4][H^+]^2}{[H_3AsO_3]}$$

$$= E^{\ominus} - 0.0592\,pH + \frac{0.0592}{2}\lg\frac{[H_3AsO_4]}{[H_3AsO_3]} \qquad (8\text{-}6)$$

The term $E^{\ominus} - 0.0592$ pH in Equation 8-6, can be considered as equal to the formal potential $E^{\ominus\prime}$ and it can be calculated from the pH value of the solution. In 0.1mol/L HCl (pH 1), $E^{\ominus\prime} = E^{\ominus} - 0.0592$. In neutral solution (pH 7), it is $E^{\ominus\prime} = E^{\ominus} - 0.41$. In strongly acid solution, H_3AsO_4 can oxidize I^- to I_2. But in neutral solution, the formal potential for As (Ⅴ)/As(Ⅲ) is 0.146V and is less than that for I_2/I^-, so the reversed reaction occurs (i.e. I_2 oxidizes H_3AsO_3 to H_3AsO_4).

1.2.2 Dependence of formal potential on complexation

If an ion in the redox couple is complexed, the concentration of the free ion in the solution is reduced. Thus the potential of the redox couple will be changed. For instance, $E^{\ominus}$ for Fe^{3+}/Fe^{2+} couple is 0.771V. In HCl solution, Fe^{3+} ion is complexed with Cl^- ion. Then the concentration of the free ion Fe^{3+} is reduced, and the potential for Fe^{3+}/Fe^{2+} is decreased. In 1mol/L HCl solution, the formal potential is 0.70V. In essence, Fe^{3+} is stabilized by complexing, that is to say, it becomes more difficult to be reduced. So the presence of complexing agents in the solution may affect the formal potential.

1.2.3 Dependence of formal potential on precipitation

Precipitation reactions involving oxidized (or reduced) ion can reduce the concentration of the free ions in the solution, so the formal potential of the redox couple is changed.

For example, the half reaction: $Cu^{2+} + e \rightleftharpoons Cu^+$, $E^{\ominus}_{Cu^{2+}/Cu^+} = 0.15V$

$I_2 + 2e \rightleftharpoons 2I^{-+}$, $E^{\ominus}_{I_2/I^-} = 0.54V$

Obviously, $E^{\ominus}_{I_2/I^-}$ is larger than $E^{\ominus}_{Cu^{2+}/Cu^+}$, so it seems that the oxidizing ability of Cu^{2+} is stronger than that of I_2 and I_2 can oxidize Cu^+. But, in fact, Cu^{2+} can oxidize I^- to virtually complete reaction because I^- reacts with Cu^+ to form precipitate CuI. It is in detail explained below.

【Example 8-1】

Calculate the formal potential for Cu^{2+}/Cu^+ couple supposing that the concentration of KI is 1mol/L.

Solution: For $Cu^{2+} + e \rightleftharpoons Cu^+$,

$$E_{Cu^{2+}/Cu^+} = E^{\ominus}_{Cu^{2+}/Cu^+} + 0.0592\lg\frac{[Cu^{2+}]}{[Cu^+]}$$

$$= E^{\ominus}_{Cu^{2+}/Cu^+} + 0.0592\lg\frac{[Cu^{2+}]}{K_{sp(CuI)}/[I^+]}$$

$$E_{Cu^{2+}/Cu^+} = E^{\ominus}_{Cu^{2+}/Cu^+} + 0.0592\lg\frac{[I^+]}{K_{sp(CuI)}} + 0.0592\lg[Cu^{2+}] \qquad (8\text{-}7)$$

When there is no other side reaction, $[Cu^{2+}] = [I^+] = 1$mol/L, $K_{sp} = 1.27\times10^{-12}$, and we can obtain the formal potential for Cu^{2+}/Cu^+ couple from Equation 8-7.

$$E^{\ominus\prime}_{Cu^{2+}/Cu^+} = E^{\ominus}_{Cu^{2+}/Cu^+} + 0.0592\lg\frac{1}{K_{sp(CuI)}}$$

$$= 0.15 + 0.0592\lg\frac{1}{1.27\times10^{-12}}$$

$$= 8.5(V)$$

Since $E^{\ominus\prime}_{Cu^{2+}/Cu^{+}}(0.85\text{V}) > E^{\ominus\prime}_{I_2/I^-}(0.54\text{V})$, then Cu^{2+} can oxidize I^- to form CuI and I_2.

2. Equilibrium constant

The equilibrium constant can be used to describe the actual extent to which a reaction will proceed toward the products.

For the redox reaction and half reactions list as below:

$$n_2\text{A (Ox)} + n_1\text{B (Red)} \rightleftharpoons n_2\text{A (Red)} + n_1\text{B(Ox)} \quad \begin{cases} \text{A(Ox)} + n_1 \rightarrow \text{A(Red)} \\ \text{B(Ox)} + n_2 \rightarrow \text{B(Red)} \end{cases}$$

$$K = \frac{[\text{A(Red)}]^{n_2}[\text{B(Ox)}]^{n_1}}{[\text{A(Ox)}]^{n_2}[\text{B(Red)}]^{n_1}}$$

$$E_{\text{A}} = E^{\ominus}_{\text{A}} + \frac{0.0592}{n_1}\lg\frac{[\text{A(Ox)}]}{[\text{A(Red)}]},\quad E_{\text{B}} = E^{\ominus}_{\text{B}} + \frac{0.0592}{n_2}\lg\frac{[\text{B(Ox)}]}{[\text{B(Red)}]}$$

After the reaction has reached equilibrium, the potentials of the two half-reactions in the solution are equal at equilibrium, $E_{\text{A}} = E_{\text{B}}$.

$$E^{\ominus}_{\text{A}} + \frac{0.0592}{n_1}\lg\frac{[\text{A(Ox)}]}{[\text{A(Red)}]} = E^{\ominus}_{\text{B}} + \frac{0.0592}{n_2}\lg\frac{[\text{B(Ox)}]}{[\text{B(Red)}]}$$

$$\lg\frac{[\text{A(Red)}]^{n_2}[\text{B(Ox)}]^{n_1}}{[\text{A(Ox)}]^{n_2}[\text{B(Red)}]^{n_1}} = \frac{n_1 n_2\left(E^{\ominus}_{\text{A}} - E^{\ominus}_{\text{B}}\right)}{0.0592} = \lg K$$

$$\lg K = \frac{n_1 n_2\left(E^{\ominus}_{\text{A}} - E^{\ominus}_{\text{B}}\right)}{0.0592} \tag{8-8}$$

If the formal potential is employed in Equation 8-8, the conditional equilibrium constant K' (条件平衡常数) is obtained.

$$\lg K' = \frac{n_1 n_2\left(E^{\ominus\prime}_{\text{A}} - E^{\ominus\prime}_{\text{B}}\right)}{0.0592} \tag{8-9}$$

We can use the conditional equilibrium constant K' to calculate the completeness of a redox reaction, and also to calculate the minimum difference between $E^{\ominus\prime}_{\text{A}}$ a nd $E^{\ominus\prime}_{\text{B}}$ (i.e. $E^{\ominus\prime}_{\text{A}} - E^{\ominus\prime}_{\text{B}}$, designed as $\Delta E^{\ominus\prime}$) needed for a quantitative reaction. The larger the value of $\Delta E^{\ominus\prime}$, the larger the K', and the more completely the redox reaction proceeds.

【Example 8-2】

How to calculate the minimum value in $\Delta E^{\ominus\prime}$ needed for a quantitative reaction?

Solution: Assume that, the number of the electrons n that is transferred in both of the half reactions is 1, and the redox reaction can be described as:

$$\text{A(Ox)} + \text{B(Red)} \rightleftharpoons \text{A(Red)} + \text{B(Ox)} \quad \begin{cases} \text{A(Ox)} + \text{e} \rightarrow \text{A(Red)} \\ \text{B(Ox)} + \text{e} \rightarrow \text{B(Red)} \end{cases}$$

Assume that, when the titration reaction is complete, less than 1/1000 of the reactant of either A(Ox) or B(Red) remains. The minimum value K' may be calculated.

$$K' = \frac{[\text{A(Red)}][\text{B(Ox)}]}{[\text{A(Ox)}][\text{B(Red)}]} = \frac{1000 \times 1000}{1 \times 1} = 10^6$$

Then we substitute $n_1 = n_2 = 1$ and $K' = 10^6$ into the Equation 8-9, we can obtain that,

$$\lg K' = \frac{E_A^{\ominus\prime} - E_B^{\ominus\prime}}{0.0592} = 6$$

$$E_A^{\ominus\prime} - E_B^{\ominus\prime} = 6 \times 0.0592 = 0.355(\text{V})$$

Therefore, under the above specific case, the value of $\Delta E^{\ominus\prime}$ must be larger than 0.355V for a quantitative redox reaction.

3. Rate of redox reactions

Electrode potentials ($E^{\ominus}$ or $E^{\ominus\prime}$ or $\Delta E^{\ominus\prime}$) will predict whether a given redox reaction can occur and calculate the degree of the completeness of the redox reactions. However, they say nothing about the kinetics and rate of the reactions, because there is no relationship between $\Delta E^{\ominus\prime}$ and reaction rate. If the rate of the electron transfer step is slow, the redox reaction may be so slow that equilibrium will be reached only after a very long time. Thus it shows that electrode potentials do not assure that a given reaction will actually effectively proceed.

The collision theory of chemical reaction and mass action law tells us, the reaction rate increases with increasing of the reactant concentrations. In addition, temperature, catalyst and acidity of the solution may also affect the rate of redox reactions.

Many redox reactions may be speeded up with temperature elevated. For example, the reaction of MnO_4^- with $C_2O_4^{2+}$ in acidic solution is rather slow at room temperature, and when the solution is heated to 75–85°C, the reaction will be fast. However, the temperature cannot be too high because part of $H_2C_2O_4$ will decompose at higher temperatures.

$$2MnO_4^- + 5C_2O_4^{2-} + 16H^+ \rightleftharpoons 2Mn^{2+} + 10CO_2\uparrow + 8H_2O$$

The rate of some redox reactions may be enhanced by catalyst. In the titration of $H_2C_2O_4$ with $KMnO_4$, the reaction is slow at beginning. With the presence of product Mn^{2+}, the reaction proceeds rapidly because Mn^{2+} can catalyze the reaction itself. So it is called self-catalyzed reaction (自催化反应).

Sometimes, one redox reaction can promote the other one.

$$2MnO_4^- + 10Cl^- + 16H^+ \rightleftharpoons 2Mn^{2+} + 5Cl_2\uparrow + 8H_2O$$

For example, the reaction of MnO_4^- with Cl^- is slow. However, it will be promoted by the redox reaction between Fe^{2+} and MnO_4^- when Fe^{2+} is present in the solution. It is called induced reaction (诱导反应). The reduction of MnO_4^- by Fe^{2+} may proceed stepwise, forming a series of intermediates such as Mn(Ⅴ), Mn(Ⅳ) and Mn(Ⅲ). These intermediates can react with Cl^-, resulting in the induced reaction. On the contrary, with addition of excess Mn^{2+}, Mn(Ⅶ) is rapidly converted to Mn(Ⅲ), along with the presence of a large amount of Mn^{2+} in the solution, the potential for Mn(Ⅲ)/Mn(Ⅱ) couple will be greatly decreased. As a result, Mn(Ⅲ) will react with Fe^{2+} but not with Cl^-. It seems that the addition of excess Mn^{2+} can prevent the oxidization of Cl^- by MnO_4^-.

PPT

Section 2 Redox Titration Curves

Like other types of titrations, a successful redox titration is based on a reaction that has a known stoichiometry between the analyte and titrant, a large equilibrium constant, and a fast reaction rate. To evaluate a redox titration, we must know the shape of its titration curve. In an acid-base titration (酸碱滴定) and a complexation titration (配位滴定), a titration curve describes the changes of the concentration of H^+ (as pH) and M^{n+} (as pM) as a function of titrant volume. For a redox titration, the potential of the redox couple is changed with the changing of the solution composition by addition of the titrant. So we can plot potential against titrant volume and construct the redox titration curve.

Let's calculate the titration curve for the titration of 20.00ml of 0.1000mol/L Fe^{2+} with 0.1000mol/L Ce^{4+} in a matrix of 1mol/L H_2SO_4. This reaction is widely used for the determination of iron in various kinds of samples.

$$Fe^{2+} + Ce^{4+} \rightleftharpoons Fe^{3+} + Ce^{3+}$$

$$Fe^{3+} + e \rightleftharpoons Fe^{2+} \qquad E^{\ominus\prime}_{Fe^{3+}/Fe^{2+}} = 0.68V$$

$$Ce^{4+} + e \rightleftharpoons Ce^{3+} \qquad E^{\ominus\prime}_{Ce^{4+}/Ce^{3+}} = 0.68V$$

$$E_{Fe^{3+}/Fe^{2+}} = E^{\ominus\prime}_{Fe^{3+}/Fe^{2+}} + 0.0592\ \lg\frac{[Fe^{3+}]}{[Fe^{2+}]}$$

$$E_{Ce^{4+}/Ce^{3+}} = E^{\ominus\prime}_{Ce^{4+}/Ce^{3+}} + 0.0592\ \lg\frac{[Ce^{4+}]}{[Ce^{3+}]}$$

The equilibrium constant for this reaction is quite large (it is approximately 7×10^{12} at 25℃). The reaction goes completion rapidly after each addition of titrant, so we assume that the titration system is at equilibrium at all times throughout the titration process. At equilibrium, the electrode potentials of all half-reactions are identical. That is, $E_{Fe^{3+}/Fe^{2+}} = E_{Ce^{4+}/Ce^{3+}} = E_{system}$ (E_{system} is the potential of the system). So the data for a titration curve can be calculated by applying Nernst equation for either Ce^{4+}/Ce^{3+} couple or Fe^{3+}/Fe^{2+} couple.

The voltage at zero titrant volume cannot be calculated, because we do not know how much Fe^{3+} is present. If $[Fe^{3+}]=0$, the calculated voltage would be $-\infty$. Factually, there must be some Fe^{3+} in the solution either as an impurity in the analyte or from oxidation of Fe^{2+} by atmospheric oxygen.

Before equivalence point (化学计量点), excess unreacted Fe^{2+} remains in solution. Therefore, we can easily calculate the concentrations of Fe^{2+} and Fe^{3+}. However, we cannot easily obtain the concentration of Ce^{4+} without solving a complicated equilibrium problem. So the potential can be calculated using Nernst equation for the Fe^{3+}/Fe^{2+} couple.

$$E_{system} = E_{Fe^{3+}/Fe^{2+}} = E^{\ominus\prime}_{Fe^{3+}/Fe^{2+}} + 0.0592\ \lg\frac{[Fe^{3+}]}{[Fe^{2+}]} = 0.68 + 0.0592\ \lg\frac{[Fe^{3+}]}{[Fe^{2+}]}$$

At equivalence point, the concentrations of Ce^{4+} and Fe^{2+} are very little and cannot be obtained from the stoichiometry (化学计量关系) of the reaction. Fortunately, equivalence-point potentials E_{eq} may be calculated because the two reactant species and the two product species have known concentration ratios at chemical equivalence.

$$E_{eq} = E^{\ominus\prime}_{Fe^{3+}/Fe^{2+}} + 0.0592\ \lg\frac{[Fe^{3+}]}{[Fe^{2+}]}$$

$$E_{eq} = E^{\ominus\prime}_{Ce^{4+}/Ce^{3+}} + 0.0592\ \lg\frac{[Ce^{4+}]}{[Ce^{3+}]}$$

$$2E_{eq} = E^{\ominus'}_{Ce^{4+}/Ce^{3+}} + E^{\ominus'}_{Fe^{3+}/Fe^{2+}} + 0.0592\ \lg\frac{[Fe^{3+}][Ce^{4+}]}{[Fe^{2+}][Ce^{3+}]}$$

At equivalence point, $[Fe^{3+}]=[Ce^{3+}]$ and $[Fe^{2+}]=[Ce^{4+}]$, therefore,

$$E_{eq} = \frac{E^{\ominus'}_{Ce^{4+}/Ce^{3+}} + E^{\ominus'}_{Fe^{3+}/Fe^{2+}}}{2} = \frac{0.68+1.44}{2} = 1.06\ (\text{V})$$

After equivalence point, the concentrations of Ce^{3+} and excess Ce^{4+} can be obtained. The potential, therefore, can be calculated using Nernst equation for Ce^{4+}/Ce^{3+} couple.

$$E_{system} = E_{Ce^{4+}/Ce^{3+}} = E^{\ominus'}_{Ce^{4+}/Ce^{3+}} + 0.0592\ \lg\frac{[Ce^{4+}]}{[Ce^{3+}]} = 1.44 + 0.0592\ \lg\frac{[Ce^{4+}]}{[Ce^{3+}]}$$

Based on the calculations above (data list in Table 8-1), we can plot the redox titration curve in Figure 8-1. It presents potential as a function of the volume of added titrant. The curve is symmetric near equivalence point because the stoichiometry is 1:1. If the stoichiometry of a redox reaction is not 1:1, the titration curve will not be symmetric about the equivalence point. In Figure 8-1, we can see that the equivalence point is marked by a steep rise in voltage (滴定突跃). For the reaction: n_2A(Ox)+ n_1B(Red) $\rightleftharpoons$ n_2A(Red) + n_1B(Ox), we assume that A(Ox) is titrant and B(Red) is the analyte. Then we can calculate equivalence-point potentials E_{eq}.

$$E_{eq} = \frac{n_1 E^{\ominus'}_{A} + n_2 E^{\ominus'}_{B}}{n_1 + n_2}$$

And we also can calculate the potentials when the titrant is 0.1% less or excessive than the amount needed for the equivalence point:

$$E^{\ominus'}_{B} + \frac{3\times 0.0592}{n_2} \sim E^{\ominus'}_{A} - \frac{3\times 0.0592}{n_1} \tag{8-10}$$

This is the potential range at the steep equivalence point break (滴定突跃范围) in titration curve. It can be seen that, the larger value of $\Delta E^{\ominus'}$ ($E^{\ominus'}_{A} - E^{\ominus'}_{B}$), the larger increase of voltage change near the equivalence point. In addition, with increasing of n_1 and n_2 numbers, the voltage change near the equivalence point can also be increased. According to the above discussion, the voltage range corresponding to the steep rise in voltage in Figure 8-1 is 0.86–1.26V.

Table 8-1 Calculation data from titration of 20.00ml of 0.1000mol/L Fe^{2+} with 0.1000mol/L Ce^{4+} in 1mol/L H_2SO_4

Volume of added Ce^{4+} (ml)	Volume percentage (%)	E (V)	Volume of added Ce^{4+} (ml)	Volume percentage (%)	E (V)
1.00	5.0	0.604	19.98	99.9	0.858
2.00	10.0	0.624	20.00	100.0	1.060
4.00	20.0	0.644	20.02	100.1	1.262
6.00	30.0	0.658	21.00	105.0	1.363
8.00	40.0	0.670	25.00	125.0	1.404
10.00	50.0	0.680	30.00	150.0	1.422
15.00	75.0	0.708	35.00	175.0	1.433
19.00	95.0	0.756	40.00	200.0	1.440

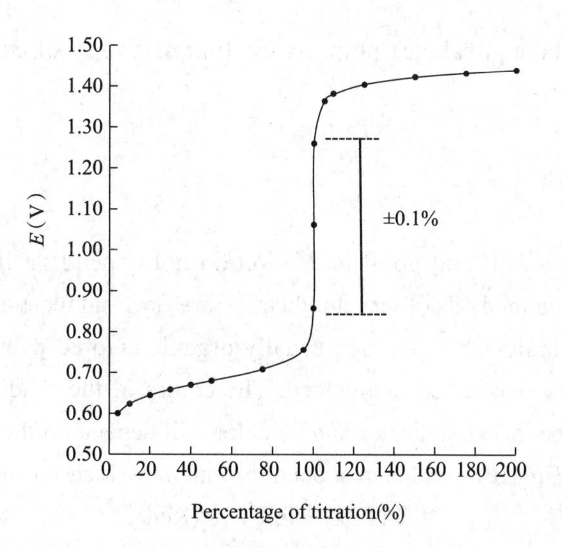

Figure 8-1 Titration curve for 20.00ml of 0.1000mol/L Fe^{2+} with 0.1000mol/L Ce^{4+} in a matrix of 1mol/L H_2SO_4

Section 3 Detection of End Point: Indicators

Similar to acid-base and complexometric titrations, it is usually convenient to use an indicator that can be observed by the naked eyes for end point detection in redox titrations. There are three methods for visual indication using self-indicator (自身指示剂), specific indicator (特殊指示剂) and redox indicator (氧化还原指示剂).

1. Self-indicator

In some situations, the color of the titrant or analyte can be directly used to monitor the progress of the redox titrations. For example, a 0.02mol/L solution of potassium permanganate ($KMnO_4$) is deep purple. A dilute solution of $KMnO_4$ with its concentration being 2×10^{-6}mol/L is pink. The product of its reduction, Mn^{2+}, is nearly colorless. During a titration with $KMnO_4$, the purple color of the MnO_4^- is removed as soon as it is added because it is reduced to colorless Mn^{2+}. As soon as the titration is complete, a fraction of a drop of excess $KMnO_4$ solution imparts a definite pink color to the solution, indicating that the reaction is complete.

2. Specific indicator

Some compounds are not oxidizing or reducing substance, but they can complex with the oxidizing or reducing reagent by color reactions. This kind of indicator is used for titrations involving iodine. Starch and iodine form a complex with a dark-blue color. The color reaction is sensitive to very small amounts of iodine. In titrations of reducing agents with iodine, the solution

remains colorless up to the equivalence point. A fraction of a drop of excess titrant of iodine turns the solution a definite blue.

3. Redox indicator

The above two methods of end-point indication do not depend on the half-reaction potentials. Examples like these first two methods of visual indication are few, and most other types of redox titrations are detected using redox indicators. They are usually organic colored dyes that are weak reducing or oxidizing agents that can be oxidized or reduced. The colors of the oxidized and reduced forms are different. The oxidation state of the indicator and its color will depend on the potential at a given point in the titration process. A half-reaction and Nernst equation can be written for the indicator:

$$\text{In (Ox)} + ne \rightleftharpoons \text{In (Red)}$$

$$E_{\text{In}} = E_{\text{In}}^{\ominus\prime} + \frac{0.0592}{n}\lg\frac{[\text{In}_{\text{Ox}}]}{[\text{In}_{\text{Red}}]} \tag{8-11}$$

The values of E_{In} and hence the ratio of $[\text{In}_{\text{ox}}]/[\text{In}_{\text{Red}}]$ are determined by the half-reaction potentials during the redox titration. This is analogous to the situation that the ratio of the different forms of a pH indicator is determined by the pH value of the solution. So the ratio, and therefore the color, will change as the potential changes during the redox titration.

If we assume, like acid-base indicators, the color of In_{Red} (reduced form of indicator) can be observed when $\frac{[\text{In}_{\text{Ox}}]}{[\text{In}_{\text{Red}}]} \leqslant \frac{1}{10}$, and the color of In_{Ox} (Oxidized form of indicator) can be observed when $\frac{[\text{In}_{\text{Ox}}]}{[\text{In}_{\text{Red}}]} \geqslant 10$. By referring to the Equation 8-11, we can obtain the potential transition range (discoloration range) for redox indicators (指示剂的变色范围),

$$E_{\text{In}}^{\ominus\prime} - \frac{0.0592}{n} \sim E_{\text{In}}^{\ominus\prime} + \frac{0.0592}{n} \tag{8-12}$$

Redox indicators for appropriate end-point detection have a transition range over a certain potential, and this transition range must overlap the steep equivalence point break of the titration curve. Additionally, the redox indicator reaction must be both rapid and reversible.

Table 8-2 Redox indicators

Indicator	$E_{\text{In}}^{\ominus\prime}$ (V) $[\text{H}^+]$ = 1mol/L	Color	
		Oxidized form	Reduced form
Nitroferroin	1.25	Pale blue	Red-violet
Ferroin	1.06	Pale blue	Red
Tris(2, 2′-bipyridine) ruthenium	1.29	Pale blue	Yellow
Tris(2, 2′-bipyridine) iron	1.12	Pale blue	Red
Diphenylbenzidine sulfonic acid	0.87	Violet	Colorless
Diphenylaminesulfonic acid	0.85	Red-violet	Colorless
Diphenylamine	0.76	Violet	Colorless
Methylene blue	0.53	Blue	Colorless
Indigo tetrasulfonate	0.36	Blue	Colorless

Oxidized form of ferroin (Pale blue) Reduced form of ferroin (Red)

Figure 8-2 Half reaction involved in ferroin with oxidized and reduced form

Table 8-2 lists some common redox indicators. Ferroin [tris(1,10-phenanthroline) iron (Ⅱ) sulfate] is one of the best indicators. It is useful for many titrations with cerium (Ⅳ). The oxidized and reduced forms of ferroin are list in Figure 8-2. It is oxidized from red to a pale blue color at the equivalence point. For this indicator, n=1 and $\Delta E^{\ominus\prime}$=1.06 can be substituted into Equation 8-12, so its potential transition range is 1.00-1.12V, which falls within the range 0.86-1.22V in the titration of Fe^{2+} with Ce^{4+} list in Figure 8-1.

Section 4 Methods Involving Iodine: Iodimetry and Iodometry

When a reducing analyte is directly titrated with iodine (to produce I^-), the method is called iodimetry (直接碘量法). In iodometry (间接碘量法), an oxidizing analyte is added to excess I^- to produce iodine. Then the produced iodine is titrated with standard thiosulfate solution.

Molecular iodine has a low solubility in water but the complex I_3^- is very soluble. So iodine solutions are prepared by dissolving I_2 in a solution of potassium iodide in which iodide is present in a large excess:

$$I_2(aq) + I^- \rightleftharpoons I_3^-$$

Therefore, I_3^- is the actual species used in the titration. A typical 0.05mol/L solution of I_3^- for titrations is prepared by dissolving 0.12mol of KI plus 0.05mol of I_2 in 1L of water. When we speak of using iodine as a titrant, we almost always mean that we are using a solution of I_2 plus excess I^-.

1. Iodimetry

Iodimetry method is often performed in neutral or mildly alkaline (pH≤8) to weakly acidic solutions. If the pH is too alkaline, I_2 will disproportionate to hypoiodate and iodide:

$$3I_2 + 6OH^- \rightleftharpoons IO_3^- + 5I^- + 3H_2O$$

There are three reasons for keeping the solution from being strongly acidic. First, the starch used for the end-point detection tends to hydrolyze or decompose in strong acid, and so the end point must be affected. Second, the reducing power of several reducing agents is increased in neutral solution. Third, the I^- produced in the reaction tends to be oxidized by dissolved oxygen in acidic solution:

$$4I^- + 4H^+ + O_2 \rightleftharpoons 2I_2 + 2H_2O$$

Because I_2 is not a strong oxidizing agent, the number of reducing agents that can be titrated is

limited. Some commonly determined substances such as S^{2-}, SO_3^{2-}, $S_2O_3^{2-}$, Sn^{2+}, AsO_3^{3-}, SbO_3^{3-} can be titrated by iodine.

2. Iodometry

When an excess of iodide is added to a solution of an oxidizing agent, I_2 is produced in an amount equivalent to the oxidizing agent. This I_2 can be titrated with a reducing agent from standard solution of sodium thiosulfate. Analysis of an oxidizing agent in this way is called an iodometric method. This indirect method can be used for the oxidizing analytes such as H_2O_2, Cu^{2+}, ClO_3^-, ClO^-, $Cr_2O_7^{2-}$, IO_3^-, BrO_3^-, MnO_4^-, NO_3^-, NO_2^- and so on. Consider, for example, the determination of dichromate:

$$Cr_2O_7^{2-} + 6I^- \text{(excess)} + 14H^+ \rightleftharpoons 2Cr^{3+} + 3I_2 + 7H_2O$$

$$I_2 + 2S_2O_3^{2-} \rightleftharpoons 2I^- + S_4O_6^{2-}$$

Each $Cr_2O_7^{2-}$ produces $3I_2$, which in turn react with $6S_2O_3^{2-}$. The millimoles of $Cr_2O_7^{2-}$ are equal to one-sixth the millimoles of $S_2O_3^{2-}$ used in the titration.

Why not titrate the oxidizing agents directly with the thiosulfate? Because strong oxidizing agents oxidize thiosulfate to oxidation states higher than that of tetrathionate (e.g., to SO_4^{2-}), but the reaction is generally not stoichiometric. Also, several oxidizing agents form mixed complexes with thiosulfate (e.g., Fe^{3+}). By reaction with iodide, the strong oxidizing agent is converted and an equivalent amount of I_2 is produced, which will react stoichiometrically with thiosulfate and for which a satisfactory indicator such as starch can be used.

Many oxidizing agents react with iodide in acid solution. The titration of the produced I_2 with thiosulfate should be performed rapidly to minimize air oxidation of the unreacted iodide. Stirring should be efficient to prevent local excesses of thiosulfate because it is decomposed in acid solution.

3. Use of Starch indicator

Starch (淀粉) is used as an indicator for iodine-based titrations. In a solution with no other colored species, it is possible to see the color of about 5μmol/L I_3^-. With starch, the limit of detection is extended.

In iodimetry (titration with I_3^-), starch can be added at the beginning of the titration. The first drop of excess I_3^- after the equivalence point causes the solution to turn dark blue.

However, in iodometry (titration of I_3^-), the starch can not be added at the beginning of the titration when the iodine concentration is high. Instead, it is added just before the end point when the dilute iodine color becomes pale yellow which indicates that most of the iodine has already been titrated with thiosulfate. If the starch is added too soon, a large amount of iodine will be adsorbed on the starch and will decrease the rate of the reaction of iodine with titrant thiosulfate, and therefore a diffuse end point would be resulted.

Starch-iodine complexation is temperature-dependent. At 50°C, the color is one tenth as intense as at 25°C. Organic solvent such as alcohols can decrease the affinity of iodine for starch and significantly reduce the sensitivity of the complexation of iodine with starch indicator. Starch indicator should be freshly prepared and used right after it is ready.

4. Preparation and standardization of solutions

4.1 Preparation and standardization of I_3^- solutions

Triiodide (I_3^-) is prepared by dissolving solid I_2 in excess KI. Sublimed I_2 is pure enough to be a primary standard, but it is seldom used as a standard because it evaporates while it is being weighed. Instead, the approximate amount is rapidly weighed, and the solution of I_3^- is standardized with a pure sample of analyte or standard solution of $Na_2S_2O_3$.

Acidic solutions of I_3^- are unstable because the excess I^- is slowly oxidized by atmospheric oxygen:

$$6I^- + O_2 + 4H^+ \rightleftharpoons 2I_3^- + 2H_2O$$

In neutral solutions, oxidation of I^- is insignificant in the absence of heat, light, and metal ions. Iodine solutions are usually standardized by a primary standard of reducing agent such as Arsenious oxide As_2O_3. Arsenious oxide is dissolved in dilute HCl or NaOH. The solution is neutralized after dissolution of As_2O_3 is complete and then arsenious acid is obtained. If arsenic (Ⅲ) solutions are to be kept for any length of time, they should be neutralized or acidified because arsenic (Ⅲ) is slowly oxidized in alkaline solution. Arsenious acid reacts with I_3^- as follows:

$$H_3AsO_3 + I_3^- + H_2O \rightleftharpoons H_3AsO_4 + 3I^- + 2H^+$$

The standardization of I_3^- with H_3AsO_3 is usually carried out at pH 7-8 in bicarbonate buffer solution. In acid solution, H_3AsO_4 may oxidize I^- to form I_2. When pH$\geqslant$11, triiodide disproportionates to hypoiodous acid, iodate, and iodide.

4.2 Preparation and standardization of sodium thiosulfate solutions

The common form of thiosulfate, $Na_2S_2O_3 \cdot 5H_2O$, is not pure enough to be a primary standard. A stable solution of $Na_2S_2O_3$ can be prepared by dissolving the reagent in high-quality, freshly boiled distilled water. Dissolved CO_2 makes the solution weakly acidic and promotes disproportionation of $S_2O_3^{2-}$:

$$S_2O_3^{2-} + H^+ \rightleftharpoons HSO_3^- + S\downarrow$$

An acidic solution of thiosulfate is unstable, but the reagent can be used to titrate I_3^- in acidic solution because the reaction of $S_2O_3^{2-}$ with I_3^- is faster than the disproportionation of $S_2O_3^{2-}$. Atmospheric oxidization of thiosulfate may be catalyzed in the presence of metal ions such Cu^{2+}. Thiosulfate solutions should be stored in the dark. Addition of 0.1g of sodium carbonate per liter keeps the pH in an optimum range for stability of the solution. Three drops of chloroform should also be added to each bottle of thiosulfate solution to help prevent bacterial growth. Sodium thiosulfate solution should be kept for 7–10 days and then be standardized on the day of their use.

Sodium thiosulfate solution is often standardized against potassium iodate KIO_3 or potassium dichromate $K_2Cr_2O_7$ which both can be used as primary standard. In the standardization, by the reaction of KIO_3 or $K_2Cr_2O_7$ with excess potassium iodide in an acidic solution, iodine is produced and immediately titrated with sodium thiosulfate solution for standardization.

$$IO_3^- + 5I^- + 6H^+ \rightleftharpoons 3I_2 + 3H_2O$$

$$Cr_2O_7^{2-} + 6I^- + 14H^+ \rightleftharpoons 3I_2 + 2Cr^{3+} + 7H_2O$$

$$2S_2O_3^{2-} + I_2 \rightleftharpoons S_4O_6^{2-} + 2I^-$$

【Example 8-3】

(1) Determination of vitamin C: iodimetry

Reducing agents can be titrated directly with standard I_3^- in the presence of starch, until reaching the intense blue starch-iodine end point. An example is the determination of vitamin C with iodimetry method: an accurately weighted sample containing vitamin C is dissolved in freshly boiled and cooled distilled water. HAc is added to make the solution mildly acidic because vitamin C can be easily oxidized in the air, especially in the basic solution. Starch is added as the indicator. Then vitamin C is immediately titrated with standard iodine solution until reaching the intense blue of starch-iodine end point.

$$\text{(vitamin C)} + I_2 \rightleftharpoons \text{(dehydroascorbic acid)} + 2HI$$

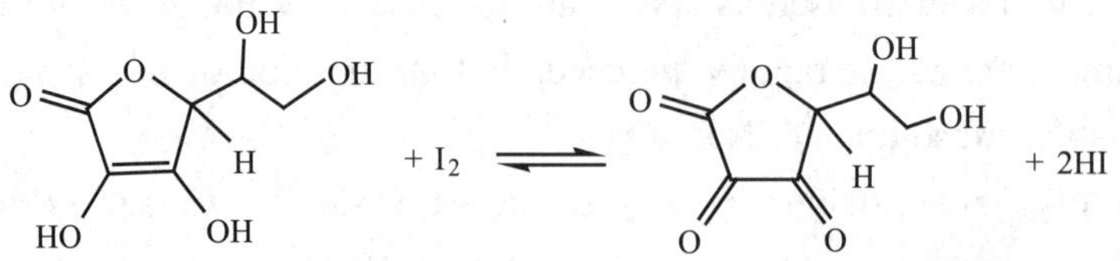

(2) Determination of copper (Ⅱ): iodometry

A 0.2000 g sample containing copper is analyzed iodometrically. Copper (Ⅱ) is reduced to copper (Ⅰ) by iodide; CuI precipitates: $2Cu^{2+} + 4I^- \rightleftharpoons 2CuI\downarrow + I_2$

If 20.00ml of 0.1000mol/L $Na_2S_2O_3$ is required for the titration of the produced I_2, what is the percent of copper in the sample?

Solution: Although there is no H^+ involved in the above redox reaction, the pH must be buffered to around 3.0–4.0. If it is too high, copper (Ⅱ) hydrolyzes and cupric hydroxide will precipitate. If it is too low, iodide can be appreciably oxidized by the air in the presence of copper as the catalyst. When the resulted iodine is titrated with thiosulfate titrant, NH_4SCN is added near the end point. Because iodine can be adsorbed on the surface of the cuprous iodide precipitate, the reaction between iodine and thiosulfate titrant proceeds very slowly. The addition of NH_4SCN makes CuI precipitate be coated with CuSCN. As a result, iodine is displaced from the surface and goes to react with $Na_2S_2O_3$. The thiocyanate should be added near the end point since it is slowly oxidized to sulfate by iodine. We can use this method to determine the copper content in copper ore, slag, plating liquid, as well as the Chinese medicine Chalcanthite.

One-half of I_2 is liberated per 1 Cu^{2+}, and 1 I_2 reacts with 1 $Na_2S_2O_3$, so each Cu^{2+} is equivalent to one $S_2O_3^{2-}$, i.e. 1mol Cu^{2+} ~ 1mol $S_2O_3^{2-}$. The molecular weight of copper is 63.54. The percent of copper in the sample is:

$$Cu\% = \frac{(CV)_{Na_2S_2O_3} \times M_{Cu}}{S} \times 100\%$$

$$Cu\% = \frac{0.1000 \times 20.00 \times 63.54 \times 10^{-3}}{0.2000} \times 100\% = 63.54\%$$

PPT

Section 5 Redox Titrations with Other Oxidants

1. Oxidization with potassium permanganate

Potassium permanganate ($KMnO_4$) is a strong and widely used oxidant with an intense violet color. Permanganate-based titrations are often carried out under strongly acidic solution such as 1mol/L H_2SO_4.

Hydrochloric acid (HCl) is not used because chloride ion is slowly oxidized by permanganate at room temperature. In strongly acidic solutions (pH≤1), $KMnO_4$ is reduced to colorless Mn^{2+}. Under this condition, potassium permanganate may act as a self-indicator for end-point detection.

$$2MnO_4^- + 5H_2O_2 + 6H^+ \rightleftharpoons 2Mn^{2+} + 5O_2\uparrow + 8H_2O \qquad E^{\ominus}_{MnO_4^-/Mn^{2+}} = 1.51V$$

In neutral or alkaline solution, the product is the brown solid, MnO_2,

$$MnO_4^- + 2H_2O + 3e \rightleftharpoons MnO_2\downarrow + 4OH^- \qquad E^{\ominus}_{MnO_4^-/MnO_2} = 0.595V$$

In strongly alkaline solution (2mol/L NaOH), the product is green manganate ion.

$$MnO_4^- + e \rightleftharpoons MnO_4^{2-} \qquad E^{\ominus}_{MnO_4^-/MnO_4^{2-}} = 0.558V$$

Potassium permanganate is not a primary standard because traces of MnO_2 are invariably present. In addition, distilled water usually contains enough organic impurities to reduce some freshly dissolved MnO_4^- to MnO_2. To prepare a 0.02mol/L stock solution, dissolve $KMnO_4$ in distilled water, boil it for an hour to accelerate the reaction between MnO_4^- and organic impurities, and allow it to stand overnight, then filter the resulting mixture through a clean, sintered-glass filter to remove precipitated MnO_2. Do not use filter paper (containing organic matter) for the filtration. Store the reagent in a dark glass bottle. Aqueous $KMnO_4$ is unstable due to the presence of the following reaction:

$$4MnO_4^- + 2H_2O \rightleftharpoons 4MnO_2\downarrow + 3O_2\uparrow + 4OH^-$$

which is slow in the absence of MnO_2, Mn^{2+}, heat, light, acids, and bases. Permanganate should be prepared and standardized for the most accurate work. Fresh dilute solutions may be prepared from 0.02mol/L stock solution, using water distilled from alkaline $KMnO_4$.

Potassium permanganate can be standardized by titration of the primary standard sodium oxalate ($Na_2C_2O_4$) or oxalic acid ($H_2C_2O_4 \cdot 2H_2O$). Dissolve dry (105°C, 2h) sodium oxalate (available as a 99.9%-99.95% pure primary standard) in 1mol/L H_2SO_4. The solution must be heated for rapid reaction. The reaction is catalyzed by the product Mn^{2+}. So it goes very slowly at first until some Mn^{2+} is formed. Then, the titration speed should be well controlled accordingly. The end point is taken as the first persistent appearance of pale pink $KMnO_4$ without fading in 30 seconds because aqueous $KMnO_4$ may slowly decompose in the air.

【Example 8-4】

(1) Determination of H_2O_2

Hydrogen peroxide can be determined with standard solution of potassium permanganate in the strongly acidic solution by adding some sulfuric acid to the solution.

$$2MnO_4^- + 5H_2O_2 + 6H^+ \rightleftharpoons 2Mn^{2+} + 5O_2\uparrow + 8H_2O$$

The above redox reaction occurs at room temperature. At beginning, MnO_4^- reacts slowly with H_2O_2, so the rate of titration is inappropriately too fast. Potassium permanganate is also the self-indicator with appearance of pale pink color of solution for the end-point detection.

$$1\text{mol } KMnO_4 \sim \frac{5}{2}\text{mol } H_2O_2$$

$$H_2O_2\% = \frac{(CV)_{KMnO_4} \times \frac{5}{2} \times M_{H_2O_2}}{V} \times 100\%$$

(2) Determination of Fe^{2+}

Fe^{2+} can be oxidized by $KMnO_4$ at room temperature.

$$MnO_4^- + 5Fe^{2+} + 8H^+ \rightleftharpoons Mn^{2+} + 5Fe^{3+} + 4H_2O$$

Phosphoric acid is added to complex the iron(III) and also decrease the potential of Fe^{3+}/Fe^{2+} couple.

Iron(Ⅱ) is not complexed. So the equilibrium of the above titration reaction shifts to the right and gives a sharp end point. An added effect of complexing the iron (Ⅲ) is that the phosphate complex is nearly colorless, while the Fe^{3+} in aqueous solution (normally present in chloride medium) is yellow. A sharper end-point color change results.

$$Fe^{2+}\% = \frac{(CV)_{KMnO_4} \times 5M_{Fe^{2+}}}{S} \times 100\%$$

2. Oxidization with potassium dichromate

In acidic solution, orange dichromate ion is a powerful oxidant that is reduced to chromic ion (Cr^{3+}):

$$Cr_2O_7^{2-} + 14H^+ + 6e \rightleftharpoons 2Cr^{3+} + 7H_2O \qquad E^{\ominus}_{Cr_2O_7^{2-}/Cr^{3+}} = 1.36V$$

Table 8-3 Formal potential of $Cr_2O_7^{2-}/Cr^{3+}$ couple in different acidic solution

Acid concentration	1mol/L HCl	3mol/L HCl	1mol/L $HClO_4$	2mol/L H_2SO_4	4mol/L H_2SO_4
$E^{\ominus\prime}_{Cr_2O_7^{2-}/Cr^{3+}}$	1.00V	1.08V	1.03V	1.11V	1.15V

Oxidation of chloride ion is not a problem with dichromate. In 1mol/L HCl, the formal potential is just 1.00V and, in 2mol/L H_2SO_4, it is 1.11V; so dichromate is a less powerful oxidizing agent than $KMnO_4$. In basic solution, $Cr_2O_7^{2-}$ is converted to yellow chromate ion CrO_4^{2-}, whose oxidizing power is nil:

$$CrO_4^{2-} + 4H_2O + 3e \rightleftharpoons Cr(OH)_3\downarrow + 5OH^- \quad E^{\ominus}_{CrO_4^{2-}/Cr^{3+}} = -0.12V$$

The great advantage of this reagent is its availability as a primary standard, and generally the solution of potassium dichromate needs not be standardized. Dichromate solutions are quite stable if they are kept in closed containers. Potassium dichromate is considered as a carcinogen and a hazardous substance for the environment. So when using this reagent, appropriate safety precaution and chemical handling procedure must be followed. Potassium dichromate is orange, but its color is not intense enough. Moreover, the green color of the product Cr^{3+} can introduce an interference to the end point detection. Therefore, potassium dichromate is not used as a self-indicator. Proper indicators of diphenylamine sulfonic acid or diphenylbenzidine sulfonic acid with distinct color changes are used to find dichromate titration end point.

【Example 8-5】

Determination of the content of soil organic matters (SOM)

The content of SOM can be measured with potassium dichromate-based method which is described as the following reactions.

$$\underset{\text{(excess)}}{2K_2Cr_2O_7} + 8H_2SO_4 + \underset{\text{(SOM)}}{3C} \xrightarrow[Ag_2SO_4]{170\text{–}180^\circ C} 2K_2SO_4 + 2Cr_2(SO_4)_3 + 3CO_2\uparrow + 8H_2O$$

$$\underset{\text{(unreacted)}}{K_2Cr_2O_7} + 6FeSO_4 + 7H_2SO_4 \rightleftharpoons K_2SO_4 + Cr_2(SO_4)_3 + 3Fe_2(SO_4)_3 + 7H_2O$$

$$Carbon\% = \frac{\frac{1}{6} \times c_{Fe^{2+}}(V_0 - V)_{Fe^{2+}} \times \frac{3}{2} \times 12.01}{S_{soil}} \times 100\%$$

In the above expression, V_0 is volume of the standard solution of $FeSO_4$ that is required for the blank solution (containing excess $K_2Cr_2O_7$ but no SOM), and V is the volume of standard solution of $FeSO_4$ that is required for unreacted $K_2Cr_2O_7$.

1g Carbon is equivalent to 1.724g SOM, therefore, SOM%=1.724×Carbon%. However, generally, 96 percent of the total amount of SOM can be oxidized by potassium dichromate. So in general, we obtain the content of SOM as:

$$SOM\% = 1.724 \times Carbon\% \times 1.04$$

3. Oxidization with Cerium(Ⅳ)

Cerium(Ⅳ) is a powerful oxidizing agent. Cerium(Ⅳ) can be used in place of $KMnO_4$ in many procedures. Its solution must be kept in acid solution (it hydrolyzes to form ceric hydroxide otherwise). The formal potential $E^{\ominus\prime}_{Cr_2O_7^{2-}/Cr^{3+}}$ can be varied with choice of the acid used (Table 8-4). Titrations are usually performed in sulfuric acid, perchloric acid or nitric acid. Sulfuric acid solutions of cerium(Ⅳ) are stable indefinitely. It seems that, the potential for Ce^{4+}/Ce^{3+} couple 1.44V is great enough to oxidize H_2O to O_2. However, reaction of cerium(Ⅳ) with water is slow. Nitric acid and perchloric acid solutions, however, do decompose slowly. Cerium(Ⅳ) solutions in HCl are unstable because Cl^- is oxidized to Cl_2, and this reaction proceeds rapidly when the solution is heated. Cerium(Ⅳ) solutions are usually prepared from diammonium ceric sulfate, $(NH_4)_4Ce(SO_4)_4 \cdot 2H_2O$ and then standardized with $Na_2C_2O_4$. Primary-standard-grade ammonium hexanitratocerate(Ⅳ), $(NH_4)_2Ce(NO_3)_6$, can be dissolved in 1mol/L H_2SO_4 and used directly without standardization.

$$Ce^{4+} + e \rightleftharpoons Ce^{3+} \quad E^{\ominus\prime}_{Ce^{4+}/Ce^{3+}} = 1.44V\ (1mol/L\ H_2SO_4)$$

Table 8-4 Formal potential of Ce^{4+}/Ce^{3+} couple in different acidic solution

Acid solution (1mol/L)	H_2SO_4	HNO_3	$HClO_4$
$E^{\ominus\prime}_{Ce^{4+}/Ce^{3+}}$	1.44V	1.61V	1.74V

Ce(Ⅳ) is yellow and Ce(Ⅲ) is colorless, but the color change is not distinct enough for cerium to be a self-indicator. Redox indicators of ferroin and other substituted phenanthroline are well suited to titrations with Ce(Ⅳ).

【Example 8-6】

Determination of malonic acid

Ce^{4+} can oxidize malonic acid to CO_2 and formic acid:

$$CH_2(COOH)_2 + 2H_2O + 6Ce^{4+} \rightarrow 2CO_2\uparrow + HCOOH + 6Ce^{3+} + 6H^+$$

This reaction can be used for quantitative analysis of malonic acid by heating a sample in 4mol/L $HClO_4$ with excess standard Ce^{4+}. Unreacted Ce^{4+} is back-titrated with standard solution of $Na_2C_2O_4$. In $HClO_4$ solution, formic acid is not oxidized by Ce^{4+}, but it can be oxidized by Ce^{4+} in 3mol/L H_2SO_4 solution. Analogous procedures are available for many alcohols, aldehydes, ketones, and carboxylic acids.

PPT

Section 6 Quantitative Calculations for Redox Titrations

The key point in the quantitative calculations for redox titrations is to accurately figure out the stoichiometric relationships between the analyte and the titrant. No matter how many kinds of reactions in the process of analysis, we can calculate the content of analyte in the sample based on the stoichiometric relationships. For example, A is the analyte. By a series of reactions from A, the substance D is obtained and titrated by titrant T. The stoichiometric relationships can be described as

$$a\text{A} \sim b\text{B} \sim c\text{C} \sim d\text{D} \sim t\text{T}$$

from which we can know the stoichiometric relationships between analyte A and titrant T, so the weight of A can be calculated by the following equation.

$$m_\text{A} = c_\text{T}V_\text{T} \times \frac{a}{t} \times M_\text{A}$$

【Example 8-7】

Analysis of bleaching powder: A 5.1225g sample of bleaching powder was dissolved in water and diluted to volume scale in 500ml volumetric flask. Then 50.00ml of this solution were treated with excess KI and a certain amount of HCl. The liberated I_2 was titrated by 0.1036mol/L $Na_2S_2O_3$ standard solution, requiring 40.16ml. Try to calculate the content of the effective chlorine in the bleaching powder.

Solution: Firstly, we can find out the reactions in the analysis process.

$$\text{CaClOCl} + 2\text{HCl} \rightleftharpoons \text{CaCl}_2 + \text{HClO} + \text{HCl}$$

$$\text{HClO} + \text{HCl} \rightleftharpoons \text{Cl}_2 + \text{H}_2\text{O}$$

$$\text{Cl}_2 + 2\text{KI} \rightleftharpoons \text{I}_2 + 2\text{KCl}$$

$$2\text{Na}_2\text{S}_2\text{O}_3 + \text{I}_2 \rightleftharpoons \text{Na}_2\text{S}_4\text{O}_6 + 2\text{NaI}$$

1mol CaClOCl ~ 1mol HClO ~ 1mol Cl_2 ~ 1mol I_2 ~ 2mol $Na_2S_2O_3$

So we know that, 1mol Cl (effective chlorine) ~ 1mol $Na_2S_2O_3$. Then the content of the effective chlorine in the sample can be calculated.

$$\text{Cl}\% = \frac{(cV)_{\text{Na}_2\text{S}_2\text{O}_3} \times M_\text{Cl}}{\frac{m_\text{S}}{500} \times 50} \times 100\%$$

$$= \frac{0.1036 \times 40.16 \times 10^{-3} \times 35.45}{\frac{5.1225}{500} \times 50} \times 100\%$$

$$=28.79\%$$

【Example 8-8】

A certain amount of primary standard KHC_2O_4 was titrated with standard solution of $KMnO_4$, requiring 16.02ml. The same amount of primary standard KHC_2O_4 as the above was neutralized by 17.05ml NaOH standard solution with concentration of 0.1196mol/L. Calculate the concentration of $KMnO_4$ standard solution.

Solution: The reactions in the analysis process are list as below.

$$2\text{MnO}_4^- + 5\text{HC}_2\text{O}_4^- + 11\text{H}^+ \rightleftharpoons 2\text{Mn}^{2+} + 10\text{CO}_2\uparrow + 8\text{H}_2\text{O}$$

$$HC_2O_4^- + OH^- \rightleftharpoons C_2O_4^{2-} + H_2O$$

According to the reactions, we can know that,

$$2mol\ KMnO_4 \sim 5mol\ KHC_2O_4;\ 1mol\ KHC_2O_4 \sim 1mol\ NaOH$$

So the stoichiometric relationship between $KMnO_4$ and NaOH is that,

$$2mol\ KMnO_4 \sim 5mol\ NaOH$$

$$\frac{(cV)_{KMnO_4}}{2} = \frac{(cV)_{NaOH}}{5}$$

$$\frac{c_{KMnO_4} \times 16.02 \times 10^{-3}}{2} = \frac{0.1196 \times 17.05 \times 10^{-3}}{5}$$

$$c_{KMnO_4} = 0.05092\ (mol/L)$$

【Example 8-9】

Determination of chemical oxygen demand (COD, 化学耗氧量): 100.00ml water sample was treated with H_2SO_4 solution and 25.00ml $K_2Cr_2O_7$ standard solution (with concentration of 0.02110mol/L) for the total oxidation of the reducing substance in the water sample. Then the unreacted $K_2Cr_2O_7$ was titrated by 0.1020mol/L $FeSO_4$ standard solution, requiring 16.18ml. Try to calculate the COD of the sample.

Solution: We can write the redox reaction between $K_2Cr_2O_7$ and $FeSO_4$.

$$K_2Cr_2O_7 + 6FeSO_4 + 7H_2SO_4 \rightleftharpoons K_2SO_4 + Cr_2(SO_4)_3 + 3Fe_2(SO_4)_3 + 7H_2O$$

COD is defined as the O_2 that is chemically equivalent to the consumed $K_2Cr_2O_7$ in the process. Each $K_2Cr_2O_7$ consumes 6e (to make $2Cr^{3-}$) and each O_2 consumes 4e (to make H_2O). So 1mol of $K_2Cr_2O_7$ is chemically equivalent to 1.5mol of O_2 for calculation.

Therefore, the stoichiometric relationships can be obtained:

$$1mol\ K_2Cr_2O_7 \sim 6mol\ FeSO_4;\quad 1mol\ K_2Cr_2O_7 \sim 1.5mol\ O_2$$

Then the amount of O_2 per liter water (C) is calculated as the result of COD measurement.

$$C = \frac{\left[(cV)_{K_2Cr_2O_7} - \frac{1}{6}(cV)_{FeSO_4}\right] \times 1.5 \times M_{O_2}}{V_{water}}$$

$$= \frac{\left(0.02110 \times 25.00 \times 10^{-3} - \frac{1}{6} \times 0.1020 \times 16.18 \times 10^{-3}\right) \times 1.5 \times 32.00}{100.00 \times 10^{-3}}$$

$= 0.1211$ (g/L)

$=121.1$ (mg/L)

重点小结

氧化还原滴定法是以氧化还原反应为基础的滴定方法，是基于电子转移的反应，反应过程比较复杂，反应速率较慢，在不同条件下的反应结果大不相同。条件电极电位考虑了氧化还原体系中各种副反应的影响，它的数值的大小，可以判断电对的实际氧化还原能力，使处理问题既简便又符合实际情况。氧化还原反应进行的程度可用条件平衡常数来衡量。氧化还原滴定可用滴定曲线表示滴定过程中待测组分浓度的变化情况。滴定曲线一般用实验方法测得，可根据 Nernst 方程式计算滴定曲线。氧化还原指示剂是一类具有弱氧化性或弱还原性的有机物质，其氧化态和还原

态具有不同的颜色。在氧化还原滴定中常用的指示剂还包括自身指示剂和特殊指示剂。

氧化还原滴定法是按所采用的滴定剂进行分类的，常用的有碘量法（又分为直接碘量法和间接碘量法）、高锰酸钾法、重铬酸钾法、铈量法等。各种方法都有其自身的特点及特定的应用范围。

氧化还原反应中反应物和产物之间的计量关系比较复杂，故计算也复杂。学习时需清楚地了解相关化学反应方程式，得到待测组分与滴定剂之间的计量关系，再根据滴定剂的浓度和体积求得待测组分的含量。

目标检测

1. Basic terminology to know

(1) Redox titration

(2) Standard potential

(3) Formal potential

(4) Conditional equivalence constant

(5) Iodimetry and Iodometry

(6) Induced reaction

2. Why is it important to have a relative large, positive value of $\Delta E^{\ominus\prime}$ for a successful redox titration?

3. How to estimate the completeness of a redox reaction?

4. For the redox reaction: $n_2A(Ox)+n_1B(Red) \rightleftharpoons n_2A(Red)+n_1B(Ox)$, $n_1=3$, $n_2=1$, what is the minimum value in $\Delta E^{\ominus\prime}$ needed for the quantitatively complete reaction?

5. It is known that, $\Delta E^{\ominus\prime}_{Fe^{3+}/Fe^{2+}} = 0.77V$, $\Delta E^{\ominus\prime}_{I_2/I^-} = 0.535V$, calculate the conditional equivalence constant K' for the reaction: $2Fe^{3+} + 2I^- \rightleftharpoons 2Fe^{2+} + I_2$

6. What are the factors that may affect the rate of the redox reaction?

7. Distinguish between iodimetry and iodometry.

8. Why are iodimetric titrations usually done in neutral solution and iodometric titrations in weak acid solution?

9. Why is iodine almost always used in a solution containing excess I^-?

10. Can you state two ways to make standard triiodide solution?

11. (a) Potassium iodate solution was prepared by dissolving 1.022g of KIO_3 in a 500ml volumetric flask. Then 50.00ml of the solution were pipetted into a flask and treated with excess KI (2g) and acid (10ml of 0.5mol/L H_2SO_4). How many moles of I_3^- are created by the reaction?

(b) The triiodide from part (a) reacted with 37.55ml of $Na_2S_2O_3$ solution. What is the concentration of the $Na_2S_2O_3$ solution?

(c) A 1.222g sample of solid containing ascorbic acid and inert ingredients was dissolved in dilute H_2SO_4 and treated with 2g of KI and 50.00ml of KIO_3 solution from part (a). Excess triiodide required 14.22ml of $Na_2S_2O_3$ solution from part (b). Find the weight percent of ascorbic acid in the sample. (The molecular weight of ascorbic acid is 176.1)

(d) Does it matter whether starch indicator is added at the beginning or near the end point in the titration in part (c)?

12. Try to design a method to determine the content of $CuSO_4$ in Chinese medicine of Chalcanthite.

13. What errors may be generated in iodimetry and iodometry? How to avoid or reduce the errors?

14. How to prepare a standard solution of Na_2SO_3?

15. What are the problems when a solution of $KMnO_4$ is prepared and standardized?

16. The content of $(NH_4)_2S_2O_8$ in a solid mixture can be determined with $KMnO_4$-based method. Peroxydisulfate sample is added to excess standard Fe^{2+} solution containing phosphoric acid for complete reduction of $(NH_4)_2S_2O_8$. Unreacted Fe^{2+} is back-titrated with $KMnO_4$ standard solution. Explain what is the purpose of phosphoric acid in the procedure?

17. The calcium in a 5.00ml serum sample is precipitated as CaC_2O_4 with ammonium oxalate. The filtered precipitate is dissolved in sulfuric acid, and the oxalate is titrated with 0.001000mol/L $KMnO_4$, requiring 4.96ml. Calculate the concentration of calcium in the serum sample.

18. Nitrite (NO_2^-) can be determined by oxidation with excess Ce^{4+}, followed by back titration of the unreacted Ce^{4+}. A sample of solid containing only $NaNO_2$ and $NaNO_3$ weighting 4.0291g was dissolved in 500.0ml. A 25.00ml sample of this solution was treated with 50.00ml of 0.1179mol/L Ce^{4+} in strong acid for 5min, and the unreacted Ce^{4+} was back-titrated with 0.04289mol/L ferrous ammonium sulfate, requiring 31.23ml to reach to the end point. Calculate the weight percent of $NaNO_2$ in the solid.

19. Chalcanthite sample 0.5190g containing $CuSO_4 \cdot 5H_2O$ was determined with iodometry method, requiring thiosulfate standard solution 18.68ml. If this thiosulfate titer is 5.15mg/ml $K_2Cr_2O_7$, calculate the content of $CuSO_4 \cdot 5H_2O$ in Chalcanthite sample.

Chapter 9　Potentiometry and Biamperometric Titration Method

学习目标

知识要求

1. 掌握　电位分析法的基本原理与基本概念；运用 Nernst 方程计算电极电位及有关离子浓度；电位滴定法和双指示电极电流滴定剂滴定终点的确定方法。

2. 熟悉　电位法中各类电极的组成、构造和测量仪器的基本性能。

3. 了解　离子选择电极的类型与应用。

能力要求

通过学习电化学分析法相关理论和知识，学会应用电位滴定法和双指示电极电流滴定法对样品溶液中的 H^+ 或者其他阴阳离子进行定量分析。

PPT

Section 1　Introduction to Electrochemical Analysis

Electrochemical analysis (电化学分析), an important part of instrument analysis methods, was established based on electrochemical principle and electrochemical properties of substances. With the multiple advantages of simple and convenient instrument, high accuracy and sensitivity, good selectivity, easy miniaturization and automation, and rapid analyzing speed, electrochemical analysis was widely employed in the field of medicine, biology, environment and materials, etc. In recent years, microelectrode and electrochemical biosensors have attracted wide attention in natural science and life science.

In electrochemical analysis, the sample solution was combined with the appropriate electrode to form an electrochemical cell and then analyzed by measuring the strength and variation of the electrical signals of electrochemical cell, such as potential, current, conductance and charge. According to the different electrical signals, electrochemical analysis methods can be divided into potentiometric analysis method (电位分析法), electrolytic analysis method (电解分析法), voltammetry (伏安法) and conductometric analysis method (电导分析法).

1. Potentiometric analysis method

It is an analytical method to determine the concentration or content of substance by measuring the functional relationship between electromotive force or electrode potential of the galvanic cell and activity (concentration) of the ion. Potentiometric analysis method includes direct potentiometric method (直接电位法) and potentiometric titration method (电位滴定法).

2. Electrolytic analysis method

It is an analytical method based on the principle of electrolysis, including electrogravimetry (电重分析法), coulometry (库仑法) and coulometric titration (库仑滴定法).

3. Voltammetry

It is an electrochemical analytical method based on the current-potential curve in the electrolysis process, including polarography (极谱法), stripping method (溶出法) and galvanometry (电流法). Galvanometry can further divide into single-indicator electrode current titration (单指示电极电流滴定法) and biamperometric titration (双指示电极电流滴定法).

4. Conductometric analysis method

It is an analytical method based on the conductivity properties of sample solution, including direct conductance method (直接电导法) and conductance titration method (电导滴定法).

This chapter will focus on potentiometric analysis method and biamperometric titration, which are the most commonly employed methods in pharmaceutical production and research.

Section 2 Basic Principles of Potentiometry

PPT

1. Electrochemical cell

Electrochemical cell is used in all electrochemical analysis methods. Electrochemical cell, consisting of two electrodes (same or different), electrolyte solution and external circuit, is a device to realize mutual conversion between chemical reaction and electrical energy. One electrode and electrolyte solution form a half-cell, and two half-cell form an electrochemical cell.

The electrochemical cell can be divided into galvanic cell (原电池) and electrolytic cell (电解池). A galvanic cell, also called a voltaic cell, uses a spontaneous chemical reaction to convert chemical energy into electric energy. However, the chemical reaction in electrolytic cell is not spontaneous, therefore additional voltage is required to convert electric energy into chemical energy. A typical galvanic cell such

as *Daniell cell* composed of copper electrode, zinc electrode, and salt bridge (盐桥) is illustrated as below (Figure 9-1).

The reactions at zinc electrode and copper electrode are represent as followed:

Zinc electrode (Anode, Negative electrode):

$$Zn \rightleftharpoons Zn^{2+} + 2e$$

Copper electrode (Cathode, Positive electrode):

$$Cu^{2+} + 2e \rightleftharpoons Cu$$

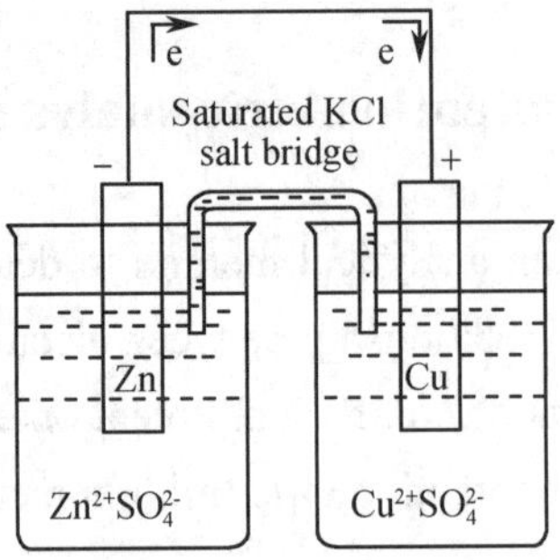

Figure 9-1 Schematic diagram of a galvanic cell that was composed of copper and zinc electrode

A line diagram of this cell can be written as follows:

$$(-)\ Zn \mid ZnSO_4\ (1mol/L) \parallel CuSO_4\ (1mol/L) \mid Cu\ (+)$$

The potential of this cell (EMF) is:

$$EMF = E_{(+)} - E_{(-)}$$

In electrochemical cell, the electrode for oxidation is anode (阳极) and the electrode for reduction is cathode (阴极). In *Daniell cell*, the zinc electrode is the anode because the oxidation of Zn occurs at this electrode, and the copper electrode is the cathode because the reduction of Cu^{2+} occurs at this electrode. At the two electrodes of *Daniell cell*, the reaction proceeds spontaneously and produces a flow of electrons from the anode to the cathode via an external conductor. It exhibits a potential of about 1.1V when no current is being drawn from it. The copper electrode is positive and the zinc electrode is negative, and it becomes a potential source of electrons to the external circuit when the cell is discharged. Galvanic cell works spontaneously, and the net reaction during discharge (called spontaneous cell reaction) for *Daniell cell* is:

$$Cu^{2+} + Zn \rightleftharpoons Cu + Zn^{2+}$$

When applying an external voltage source (外加电压) with its potential somewhat larger than 1.1V, and connecting its positive terminal to the copper electrode and its negative terminal to the zinc electrode, the cell can work as an electrolytic cell. Electrons flow from the negative terminal to the zinc electrode at which Zn^{2+} is reduced. Cu is oxidized at the copper electrode which produces electrons flowing to the positive terminal of the voltage source. Then zinc electrode is forced to be the anode and copper electrode is forced to be the cathode. The net reaction occurs as:

$$Cu + Zn^{2+} \rightleftharpoons Cu^{2+} + Zn$$

A line diagram of electrolytic cell can be written as follows:

$$(+)\ Cu \mid CuSO_4\ (1mol/L) \parallel ZnSO_4\ (1mol/L) \mid Zn\ (-)$$

By convention, in the line diagram of a cell, a single vertical line indicates a phase boundary (or interface) between an electrode phase and a solution phase or two solution phases. The double vertical lines indicate two-phase boundaries. In the cells list above, the double line represents the liquid junction between two different solutions. And there is a potential associated with the liquid junction, called the liquid junction potential (液接电位, E_{lj}). It is usually in the form of a salt bridge which can prevent mixing of the two solutions.

2. Liquid junction potential

Liquid junction potential (液接电位, E_{lj}) results from the unequal diffusion of the ions on each side of the boundary between any two different solutions. The two solutions can be different in the composition or can be the same in composition but different in ion concentration.

For example, solutions of 0.1mol/L HCl and 0.01mol/L HCl separated by a porous membrane (Figure 9-2A). Since the concentration of HCl on the right side of the membrane is greater than that on the left side of the membrane, there is a net diffusion of H^+ and Cl^- in the direction of the arrows. However, the mobility of H^+, is greater than that for Cl^-, as shown by the difference in the lengths of their respective arrows. As a result, the solution on the left side of the membrane develops an excess of H^+ and has a positive charge. Simultaneously, the solution on the right side of the membrane develops a negative charge due to the greater concentration of Cl^-. Hence, a constant potential difference (*i.e.* liquid junction potential) is quickly achieved between the two solutions.

The magnitude of the liquid junction potential is determined by the ionic composition of the solutions on the two sides of the interface and may be as large as 30–40mV. When the potential of an electrochemical cell is measured, the contribution of the liquid junction potential must be included. Thus, equation is rewritten as

$$\text{EMF} = E_{(+)} - E_{(-)} + E_{lj}$$

The magnitude of a salt bridge's liquid junction potential can be minimized by using a salt, such as KCl, since the mobilities of the cation and anion are approximately equal. A saturated solution of potassium chloride, KCl, is the electrolyte that is most widely used. This electrolyte can reduce the junction potential to a few millivolts or less. For this reason salt bridges are frequently constructed using solutions that are saturated with KCl (Figure 9-2B).

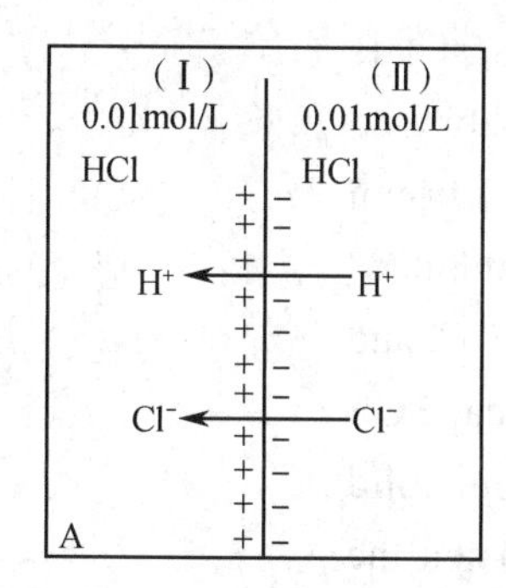

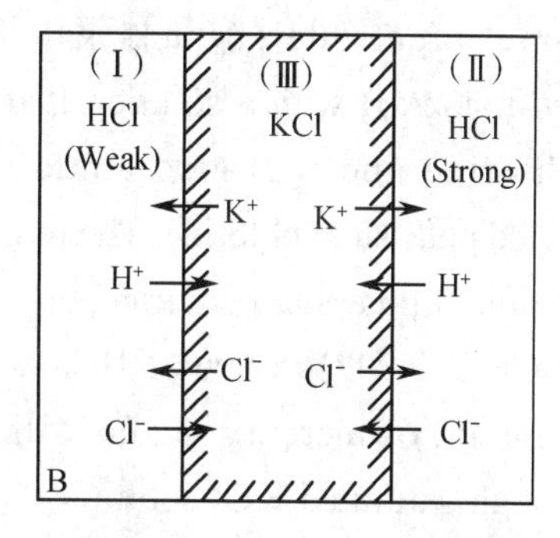

Figure 9-2 Schematic diagram of formation and elimination of liquid junction potential

Section 3 Reference Electrode and Indicator Electrode

Electrodes are divided into reference electrodes (参比电极) and indicator electrodes (指示电极) according to their role in chemical cells. A reference electrode is a half-cell having a known electrode potential at constant temperature and is independent on the composition of the analyte (分析物) solution. An indicator electrode is immersed in a solution of the analyte and develops a potential that depends on the activity of the analyte.

1. Reference electrode

The ideal reference electrode has a potential that is accurately known, constant, and completely insensitive to the concentration of the analyte solution. As a reference electrode, it should have the following basic requirements: (1) Constant potential, (2) Good reproducibility, (3) Simple instrument, (4) Convenient and durable use.

Recently, the most widely used reference electrode is saturated calomel electrode (饱和甘汞电极，SCE) and silver/silver chloride electrodes (银-氯化银电极，SSE).

1.1 Standard hydrogen electrode

The standard hydrogen electrode (标准氢电势, SHE) is an important reference electrode used to establish standard-state potentials for other half-reactions. However, a SHE is seldom used because it is somewhat troublesome to maintain and use. The SHE consists of a Pt electrode immersed in a solution in which the hydrogen ion activity is 1.00 and H_2 gas is bubbled at a pressure of 1 atm. The hydrogen electrode in Figure 9-3 can be represented as:

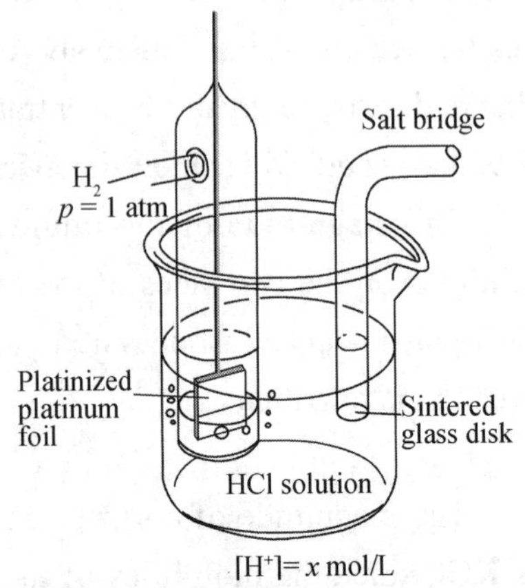

Figure 9-3 The hydrogen gas electrode

$$Pt(s), H_2\ (g, 1\ atm)\ |\ H^+ (aq, a = 1.00)\ ||$$

1.2 Saturated calomel electrode

Saturated calomel electrode (饱和甘汞电极, SCE) is composed of mercury in contact with a solution that is saturated with mercury (I) chloride and that also contains a known concentration of saturated potassium chloride. The term "saturated" refers to the concentration of potassium chloride; and at 25°C, and the potential of the SCE is 0.242V versus NHE. A typical SCE consists of a small amount of mercury mixed with some solid Hg_2Cl_2 (calomel), enough saturated KCl solution to moisten the Hg/Hg_2Cl_2 mixture and solid KCl to maintain saturation (Figure 9-4). A small hole connects the two tubes, and an asbestos fiber serves as a salt bridge to the solution in which the SCE is immersed. The stopper in the outer tube may be removed when additional saturated KCl is needed.

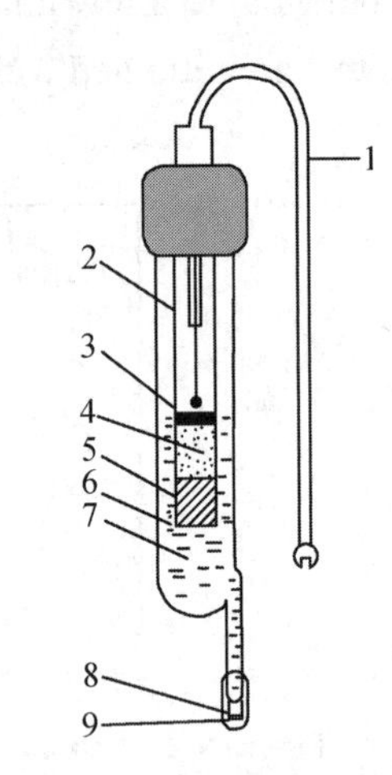

Figure 9-4 The saturated calomel electrode

1. Electrode line; 2. Glass tube; 3. Mercury; 4. A plaste mixture of Hg and Hg_2Cl_2; 5. Asbestos or pulp; 6. Outer Glass tube; 7. Saturated KCl solution; 8. Porous ceramic junction; 9. Rubber stopper

Calomel half-cells: $Hg|Hg_2Cl_2, KCl(a)$

Half-reaction: $Hg_2Cl_2 + 2e \rightleftharpoons 2Hg + 2Cl^-$

The equation of the electrode potential (25°C): $E = E^{\ominus}_{Hg_2Cl_2/Hg} - 0.0592\lg a_{Cl^-}$ (9-1)

According to Equation 9-1, the standard potential is dependent on the concentration of KCl as illustrated in Table 9-1.

Table 9-1 The electrode potential of calomel electrode

c_{KCl} (mol/L)	≥3.5 (Saturated)	1	0.1
Electrode potential (V)/ 25°C	0.2412	0.2801	0.3337
The relationship between Electrode potential (E) and Temperature (T)	E=0.2412−6.61×10^{-4}(T−25)−1.75×10^{-6}(T−25)^2		

1.3 Bis-salt bridge saturated calomel electrode

See the structure of bis-salt bridge SCE (双液接饱和甘汞电极, bi-SCE) in Figure 9-5. An outer glass tube is connected at the lower end of SCE and filled with appropriate electrolyte solution (commonly KNO_3). Bi-SCE should be used, when SCE is unable to satisfy the demand.

(1) KCl in SCE can react with ions in sample solution. For example, Cl^- ions in SCE can react with Ag^+ ions to produce AgCl sediment if SCE is used to detect Ag^+.

(2) KCl in SCE will penetrate into the sample solution and cause interference if the measured ion is Cl^-or K^+.

(3) When the sample solution contains I^-、CN^-、Hg^{2+} or S^{2-}, the potential of SCE will slowly and orderly change (drift) with time, and even destroy the function of SCE electrode in serious cases

(4) The residual E_{lj} between SCE and sample solution is large and unstable. For example, bi-SCE is often used in non-aqueous titration.

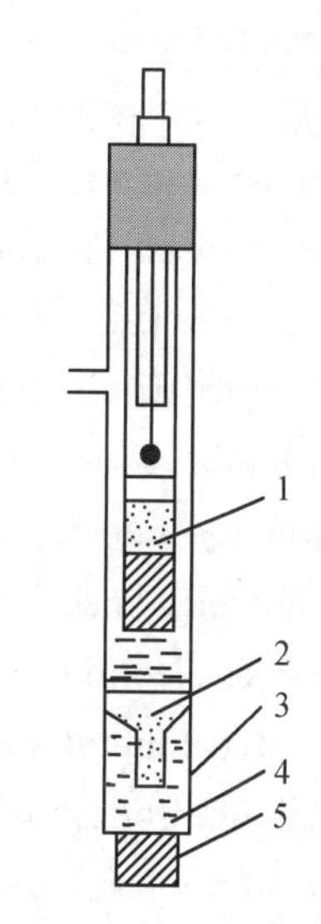

Figure 9-5 The bis-salt bridge saturated calomel electrode

1. Saturated calomel electrode; 2. Ground-glass joint; 3. Outer glass tube; 4. KNO_3 solution; 5. Porous ceramic junction

1.4 Silver/Silver chloride electrodes

Silver/silver chloride electrode (银-氯化银电极, SSE) consists of a silver electrode immersed in a solution of potassium chloride that has been saturated with silver chloride. It is usually used as internal reference electrode of ion-selective electrode (离子选择性电极, ISE).

Half-cell electrode: Ag | AgCl, KCl(a)

Half-reaction: AgCl + e = Ag + Cl^-

The equation of the electrode potential (25°C): $E = E^{\ominus}_{AgCl/Ag} - 0.0592\lg a_{Cl^-}$ (9-2)

The potential of Ag/AgCl electrode is determined by the activity (or concentration) of Cl^-. If the activity of Cl^- and temperature are constant, the potential of Ag/AgCl electrode unchanged. The potentials of Ag/AgCl for these electrodes under different concentration of Cl^- are given in Table 9-2.

Table 9-2 The potential of SSE vs SHE

c_{KCl} (mol/L)	≥3.5 (Saturated)	1	0.1
Potential (V)/25°C	0.1990	0.2223	0.2880

2. Indicator electrode

An ideal indicator electrode responds rapidly and reproducibly to changes in the concentration of an analyte ion (or group of analyte ions). The indicating electrode meets the following requirements:

(1) The relationship between the electrode potential and analyte's activity (concentration) can be expressed by Nernst equation;

(2) Owning Rapid respond, wide linear range and well reproducibility;

(3) High selectivity to the analyte;

(4) Simple structure and convenient to use.

Frequently-used indicator electrodes are of two types: metallic electrode (金属基电极) and ion-sensitive electrode (ISE, 离子选择电极).

2.1 Metallic indicator electrode

The potential of a metallic electrode is determined by the position of a redox reaction at the electrode–solution interface. Metallic electrodes are commonly used in potentiometry, including metal/metal ion electrode, metal/insoluble metal salt electrode, metal/insoluble metal oxide electrode, metal/metal complexes electrode, and noble electrode, etc.

2.1.1 Metal/metal ion electrode

A metal electrode in contact with a solution containing the cation of the same metal is called metallic-metallic ion indicator electrode (金属-金属离子指示电极), or electrodes of the first kind, 第一类电极). The potential of the electrode is determined by the concentration of metal ion. The electrode system can be represented by M/M^{n+}, in which the line represents an electrode–solution interface. Ag, Cu, Hg, and Zn are considered as electrodes of this type.

For example: $Ag\text{-}Ag^+$ indicator electrode: $Ag \mid Ag^+(a)$

Half-reaction: $Ag^+ + e = Ag$

The equation of the electrode potential (25°C): $E = E^{\ominus}_{Ag^+/Ag} + 0.0592\lg a_{Ag^+}$ (9-3)

2.1.2 Metal/insoluble metal salt electrode

Metal/insoluble metal salt indicator electrode (金属-金属难溶盐电极) is composed of a metal electrode (M) coated with the same metal salt (M_mX_n) and inserted into the anionic solution (X^{m-}) of M_mX_n. The electrode potential can be used as an indicator electrode for the determination of the activity (or concentration) of the anions. The general form can be expressed as: $M \mid M_mX_n \mid X^{m-}$, where M_mX_n is a slightly soluble salt. This kind electrode is also called electrodes of the second kind (第二类电极) because it has two phase interfaces (相界面).

For example: Ag-AgCl electrode: $Ag \mid AgCl, KCl\ (a)$

Half-reaction: $AgCl + e \rightleftharpoons Ag + Cl^-$

The equation of the electrode potential (25°C): $E = E^{\ominus}_{AgCl/Ag} - 0.0592\lg a_{Cl^-}$ (9-4)

Note that, as the activity of chloride increases, the potential decreases.

2.1.3 Metal/insoluble metal oxide electrode

Metal/insoluble Metal oxide electrode (金属-金属难溶氧化物电极) is composed of a metal electrode (M) and metal oxide (M_mO_n) that also belong to electrodes of the second kind. For example, Sb coated with Sb_2O_3 is inserted H^+ solution to obtain Sb/Sb_2O_3 electrode.

Sb/Sb_2O_3 electrode: $Sb \mid Sb_2O_3, H^+\ (a)$

Half-reaction: $Sb_2O_3 + 6H^+ + 6e \rightleftharpoons 2Sb+3H_2O$

The equation of the electrode potential (25°C):

$$E = E^{\ominus}_{Sb_2O_3/Sb} + 0.0592\lg a_{H^+} = E^{\ominus}_{Sb_2O_3/Sb} - 0.0592\text{pH} \tag{9-5}$$

From Equation 9-5, Sb/Sb_2O_3 electrode is an indicator electrode. Since Sb_2O_3 can be dissolved in strong acids and weak bases solution, Sb/Sb_2O_3 electrode is suitable to use in solution with pH ranging from 3 to 12.

2.1.4 Metal/metal complexes electrode

Metal/metal complexes electrode (金属-金属配合物电极) for EDTA is constructed by coupling a Hg^{2+}/Hg electrode to EDTA by taking advantage of its formation of a stable complex with Hg^{2+} based on complexation reactions. At the same time, there is another metal ion (M^{n+}) in the sample solution. Moreover, the equilibrium constant (平衡常数) of MY is lower than HgY. For example, Hg/Hg-EDTA electrode is inserted into a sample solution containing Ca^{2+}.

Half-cell: Ca^{2+}, Hg | HgY^{2-} (a_1), CaY^{2-} (a_2), Ca^{2+}(a_3)

Half-reaction: $Hg^{2+} + 2e \rightleftharpoons Hg$ $\quad Hg^{2+} + Y^{4-} \rightleftharpoons HgY^{2-}$ $\quad Ca^{2+} + Y^{4-} \rightleftharpoons CaY^{2-}$

The equation of the electrode potential (25°C)

$$E = E^{\ominus}_{Hg^{2+}/Hg} + \frac{0.0592}{2}\lg a_{Hg^{2+}} \tag{9-6}$$

By the coordination equilibrium, we can conclude that

$$E = E^{\ominus}_{Hg^{2+}/Hg} + \frac{0.0592}{2}\lg\frac{K_{CaY^{2-}}a_{HgY^{2-}}}{K_{HgY^{2-}}a_{CaY^{2-}}} + \frac{0.0592}{2}\lg a_{Ca^{2+}} \tag{9-7}$$

Hg/Hg-EDTA electrode is usually used as an indicator electrode and SCE as a reference electrode in titration of Ca^{2+} with EDTA. When the titration reaction is close to stoichiometric point (化学计量点)，$K_{CaY^{2-}}$ and $K_{HgY^{2-}}$ are both constant. HgY^{2-} is so stable that its concentration remain still. Therefore, Equation 9-7 can be shortly written as Equation 9-8. This kind electrode is also called electrodes of the third kind because involve in three chemical equilibrium. Metal/metal complexes electrode can be used to detect more than thirty metal ion in coordination reaction with EDTA.

$$E = K' + \frac{0.0592}{2}\lg a_{Ca^{2+}} \tag{9-8}$$

2.1.5 Noble electrode

Noble electrode (惰性电极) is produce by inserting noble metal, such as Pt or Au, into the solution containing different redox electric pairs. For example,

Half-cell: Pt | Fe^{3+}, Fe^{2+}

Half-reaction: $Fe^{3+} + e \rightleftharpoons Fe^{2+}$

The equation of the electrode potential (25°C):

$$E = E^{\ominus}_{Fe^{3+}/Fe^{2+}} + 0.0592\lg\frac{a_{Fe^{3+}}}{a_{Fe^{2+}}} \tag{9-9}$$

In this kind of electrodes, noble metal is only used as an indicator electrode to determine redox reaction in homogeneous phase solution and does not react on their own, the function of which is only used to transfer electrons.

2.2 Ion-selective electrode

Ion-selective electrode (离子选择电极, ISE), also call membrane electrode, is a thin membrane capable of responding selectively to the specific ion. There are no charge transfer and half-cell reaction on ion-selective electrode. The mechanism of ISE is based on diffusion and exchange of analyzed ions on the surface of membrane electrode. Moreover, the relationship between the potential of ISE and the activity (concentration) of analyzed ions can be expressed in Nernst equation.

$$E_{ISE} = K \pm \frac{2.303RT}{nF} \lg a_i \tag{9-10}$$

Where K is a constant; a_i is the activity of the analyzed ions. "+" and "–" are chose, respectively, corresponding to cation and anion.

Ion-selective electrodes is highly selective and sensitive that are the widely used method for the determination of pH, Na^+, K^+, and Cl^-, etc.

3. Combination electrode

Combination electrode is an electrode integrated indicator electrode with reference electrode. The commonly used pH electrodes consist of Ag/AgCl and Hg/Hg_2Cl_2 electrodes.

Section 4 Direct Potentiometry

The potential of a galvanic cell can be measured by immersing a proper reference electrode and an indicator electrode into the analyte solution. According to the Nernst equation, the activity of the analyte can be calculated, which is called direct potentiometry. Direct potentiometry is widely used to detect pH and the activity of cation and anion for being rapid, efficient and convenient.

1. Determination of hydrogen ion activity

Potentiometry measurement usually employs SCE as a reference electrode, hydrogen, quinone-hydroquinone, antimony or glass electrode as an indicator electrode. Among these indicator electrodes, glass electrode is most commonly used.

1.1 pH glass electrode

1.1.1 The composition and structure of glass membranes

pH glass electrode is the earliest developed electrode. Figure 9-6 shows a typical cell for measuring pH using pH glass electrode. The spherical glass membrane (球形玻璃膜) with the thickness of 0.1mm is pH-sensitive. Buffer solution containing 0.1mol/L HCl or KCl is used as internal reference solution, and Ag/AgCl electrode is used as internal reference electrode. Furthermore, highly insulated conductors and

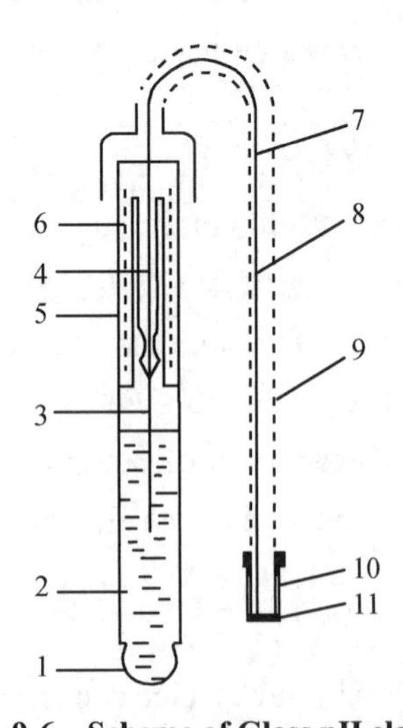

Figure 9-6 Scheme of Glass pH electrode.
1. Thin pH glass membrane; 2. Buffer solution; 3. Internal Ag-AgCl reference electrode; 4. Electrode leg wire; 5. Glass tube; 6. Electrostatic isolation layer; 7. Electrode conducting wire; 8. Insulation plastic; 9. Metal isolation cover; 10. High insulation plastic; 11. Electrode joint

leads are used at the top of the electrode, and metal shielding layer is used outside the conductors and leads wires to avoid electric leakage and electrostatic interference induced by the high internal resistance (内阻, >100MΩ) of the electrode. It is noted that thin pH glass membrane is fragile so that it is necessary to protect the membrane when using pH glass electrode.

1.1.2 Response mechanism of pH glass electrode

The selective response of glass electrode to H^+ is dependent on the composition of the membrane. pH glass electrode often consists of approximately 72.2% SiO_2, 21.4% Na_2O, and 6.4% CaO. Membrane potential generation contain three main steps, including hydration (水化) of glass membrane, ion exchange and diffusion.

Before use, pH glass electrode is immersed into water for a period of time to form a swelling hydrated silica gel layer with the thickness of 10^{-5}-10^{-4}mm. Then, an ion-exchange reaction between singly charged cations in the interstices of the glass lattice and hydrogen ions from the solution (Figure 9-7). The ion-exchange reaction can then be written as

$$H^+(aq) + Na^+Gl^-(\text{glass membrane}) \rightleftharpoons Na^+(aq) + H^+Gl^-(\text{glass membrane})$$

The reaction equilibrium constant is large, so H^+ replace all Na^+ on the membrane surface. However, the amount of H^+ increases with the increase in depth of gel layer, which is contrary to Na^+. In the center of gel layer, there is no ion exchange and Na^+ occupy all sites.

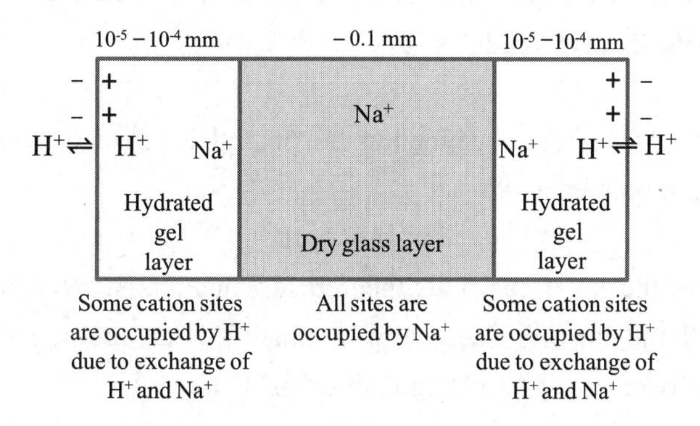

Figure 9-7 Hydration layer of pH glass electrode

The fully hydrated glass electrode is put into the sample solution. Ion diffusion appears because of the difference of H^+ between hydration layer and sample solution. H^+ diffusion cause the charge distribution change on the membrane surface and two-phase interface, leading to the generation of double electric layer. When the diffusion reaches equilibrium, boundary potential (相界电位, E_b) is constant, as well as internal boundary potential (内相界电位, E_{ib}). Apparently, boundary potential is dependent on the activity of H^+ between two phases.

$$E_b = K_1 + \frac{2.303RT}{F}\lg\frac{a_o}{a'_o} \quad (9\text{-}11)$$

$$E_{ib} = K_2 + \frac{2.303RT}{F}\lg\frac{a_i}{a'_i} \quad (9\text{-}12)$$

Where a_o and a_i are the activities of H^+ outside and inside the membrane, a'_o and a'_i are the activities of H^+ on the outer surface of membrane and the inner surface of hydration layer, K_1 and K_2 is the constant related to the physical properties of inner and outer glass membrane surface.

Membrane potential can be written as

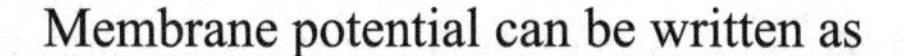

$$E_m = E_b - E_{ib} = \left(K_1 + \frac{2.303RT}{F}\lg\frac{a_o}{a_o'}\right) - \left(K_2 + \frac{2.303RT}{F}\lg\frac{a_i}{a_i'}\right) \tag{9-13}$$

For the same pH electrode, $K_1 = K_2$, and $a_o' = a_i'$, therefore

$$E_m = \frac{2.303RT}{F}\lg\frac{a_o}{a_o'} \tag{9-14}$$

Usually, pH of the reference solution is constant so that a_i is also constant, thus

$$E_m = K' + \frac{2.303RT}{F}\lg a_o \tag{9-15}$$

For whole glass pH electrode, the total potential E_{glass} is the sum of membrane potential (E_m) and internal reference electrode potential (E_{ir}) so that the potential of pH glass electrode is related with the activity of H^+. The equation can be written as:

$$E_{glass} = E_{ir} + E_m = K + \frac{2.303RT}{F}\lg a_o = K - \frac{2.303RT}{F}\text{pH} \tag{9-16}$$

$$\left(K = E_{ir} - \frac{2.303RT}{F}\lg a_i\right)$$

Where K is called electrode constant, and the equation of E_{glass} indicates linear relationship between the potential of glass pH electrode and pH value of the solution outside the membrane.

1.1.3 Function of pH glass electrode

(1) Conversion coefficient

Conversion coefficient (换算系数) is defined as the potential difference caused by one unit change of pH value of solution and expressed as S

$$S = -\Delta E/\Delta \text{pH} \tag{9-17}$$

Where S is the slope of the E-pH curve, its theoretical value is $2.303RT/F$ that means the potential will change by 59.2mV with the one unit change of pH value. With the prolonged use, glass pH electrode should be discarded when S decrease to 52mV/one unit of pH value.

(2) Alkali and acidic error

Alkali error (碱差), also called sodium error, refers to the negative difference of pH between the detect value and the true value in alkali solution (pH>9). When pH values of the solution are above 9, H^+ concentration is very small compared with other ions. The electrode becomes appreciably response to the other ions such as Na^+, K^+ and so on. In effect, the electrode appears to "see" more hydrogen ions than the real actual hydrogen ion in the solution. Therefore, the pH reading is lower than the true pH.

Acidic error (酸差), refers to the positive difference of pH between the detect value and the true value in acidic solution (pH<1). There are several contributors to the acid error, but not all the causes and effects can be well explained. One of the causes is that the solution with pH less than 1 has a low activity of H_2O so that H^+ cannot easily diffuse into hydration layer since H^+ transfers by the formation of H_3O^+. So the pH reading is higher than the true pH.

Therefore, glass pH electrode usually is used under the pH value ranged from 1 to 9.

(3) Asymmetric potential

Ideally, asymmetric potential (不对称电位，E_{asym}) should in principle be zero when the concentrations of analyte on both sides of the membrane are equal. Frequently, the term E_{asym} accounts for the fact that the membrane potential is usually not zero under these conditions. The sources of the asymmetry potential are unclear but undoubtedly containing factors as differences in strain on the two surfaces of the membrane created during manufacture, mechanical abrasion on the outer surface during use, and chemical etching of the outer surface. To eliminate the bias caused by the asymmetry potential, all membrane electrodes must be activated in pure water for at least 24 hours and calibrated against one or more standard analyte solutions. Calibrations should be carried out at least daily and more often when the electrode is heavily used. After use, the membrane electrodes should be stored in 3mol/L KCl solution.

(4) Internal electrode resistance

Internal electrode resistance (电极内阻) of glass electrode is usually large, ranging from 50 to 500MΩ so that only small current is allowed to pass through to produce high error when glass electrode is used to detect cell potential. Internal electrode resistance increases with the prolonged time, leading to reduced sensitivity. Thus, glass electrode should be altered after long-term use.

1.2 Measurement principles and method

1.2.1 Measurement principles

In potentiometry, a galvanic cell is composed of glass pH electrode as an indicator electrode and SCE as a reference electrode. It can be written as:

(–) Ag, AgCl |HCl(a) | glass membrane | sample solution (a_{H^+}) || KCl (sat.)| Hg_2Cl_2, Hg (+)

Associated with Equation 9-16, the potential equation is obtained:

$$E = E_{SCE} - E_{glass} = E_{SCE} - \left(K - \frac{2.303RT}{F}\text{pH}\right) = K' + \frac{2.303RT}{F}\text{pH} \tag{9-18}$$

$$\text{At } 25°\text{C}, \quad E = K' + 0.0592\text{pH} \tag{9-19}$$

Therefore, pH value can be calculated by the cell potential according to the linear relationship between the potential and pH.

1.2.2 Methods

In Equation 9-19, K' is a combination of the several potential affected by electrode constant, the composition of the sample solution, operating time and so on. It is difficult to detect and calculate. Therefore, pH of standard solution with known concentration (pH_s) and sample solution (pH_x) with unknown concentration were determined under the same condition.

$$E_s = K' + 0.0592\text{pH}_s \tag{9-20}$$

$$E_x = K' + 0.0592\text{pH}_x \tag{9-21}$$

According to Equation 9-20 and 9-21, we conclude that pH_x (Equation 9-22) can be obtained via the determination of E_s and E_x.

$$\text{pH}_x = \text{pH}_s + \frac{E_x - E_s}{0.0592} \tag{9-22}$$

2. Determination of activity (concentration) of other anions and cation

2.1 Basic structure and electrode potential of ion-selective electrode

Ion-selective electrode (ISE) is a kind of electrodes selectively toward a given ion or ions. Ion-selective electrode consists of glass membrane, electrode tube (support), internal reference electrode and internal reference solution (Figure 9-8). Glass membrane and internal reference electrode both contain as

same analyzed ions as the sample solution (contain the analyzed ions as same as those in the sample solution). When the glass membrane is immersed into sample solution, the ions with selective response inside and outside the membrane establish potential difference on both sides of the membrane through ion exchange or diffusion, and form membrane potential after balance. Similar as glass pH electrode, the potential of ion-selective electrode is only related with the activity of responsive ions when the composition of internal reference solution remained unchanged. Their relationship can be expressed as Nernst equation:

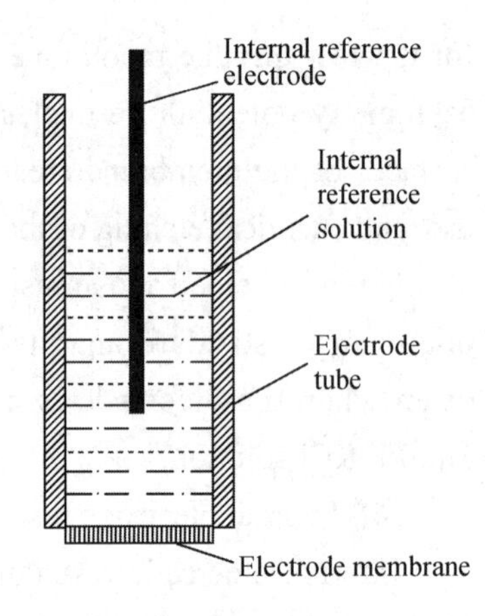

Figure 9-8 Ion-selective electrodes

$$E_{\mathrm{ISE}} = K \pm \frac{2.303RT}{nF} \lg a_{\mathrm{i}} \tag{9-23}$$

It should be noted that the electrode potential of some ion-selective electrodes is not only established by ion exchange and diffusion, but also related to ion association and coordination; the mechanism of some ion-selective electrodes is still not very clear.

2.2 Ion-selective electrode classification and common electrodes

2.2.1 Ion-selective electrode classification

According to IUPAC's recommendations on the naming and classification of ion selective electrodes, the names and classifications of ion selective electrodes are as follow:

(1) Primary electrode

The primary electrode is an ion selective electrode that the electrode membrane directly contacts with the test solution. According to the different materials of electrode membrane, it can be divided into crystalline electrode (晶体电极) and non-crystalline electrode (非晶体电极).

Crystalline electrode means that the membrane material is made of crystal of one or several compounds. According to the preparation method of electrode membrane, crystalline electrode is divided into homogeneous membrane electrode and heterogeneous membrane electrode. The membrane material of the homogeneous membrane electrode is made of insoluble single crystal, multi-crystal or mixed crystal. The membrane material of the heterogeneous membrane electrode is added with inert materials (such as silicone rubber, PVC or paraffin), except for electroactive substance.

Non-crystalline electrode means that the membrane material is made of amorphous materials or compounds uniformly dispersed in inert supports. Among them, the glass electrode which is produced by specific glass is called rigid matrix electrode (刚性基质电极), such as pH glass electrode, sodium electrode, potassium electrode, lithium electrode, etc. Electrode with a mobile carrier is another kind of amorphous electrode, also known as liquid membrane electrode. Its electrode membrane is made of inert porous membrane impregnated with some liquid ion exchanger or neutral carrier. Mobile carrier electrode is a kind of ion selective electrode which is widely used in pharmacy and inspection, such as some commercial Li^+, K^+, Na^+, Ca^{2+}, Mg^{2+}, Ba^{2+}, Cd^{2+} selective electrodes.

(2) Sensitized ion-selective electrode

Through the interface reaction, the analyte can be transformed into the ion that can be measured by the basic electrode to realize the indirect determination of the analyte. According to the different properties of the interface reaction, typical sensitized ion-selective electrode can be divided into ammonia sensitive electrode and urease electrode.

2.2.2 Commonly used electrodes

Fluoride ion selective electrode (氟离子选择电极) is made of single crystal of lanthanum fluoride and sealed at one end of the electrode tube. The tube is filled with 0.1mol/L NaF and 0.1mol/L NaCl solution as internal electrolyte (Figure 9-9). The Ag-AgCl electrode is used as the internal reference electrode. F^- can diffuse in single crystal of lanthanum fluoride that the electrode potential is related to the activity of F^-, which accords with Nernst equation.

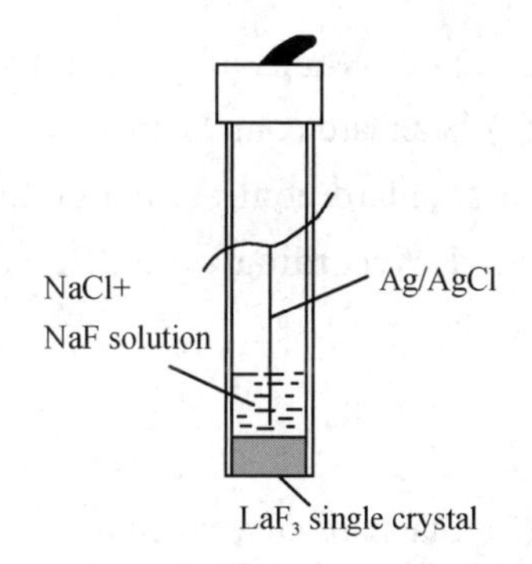

Figure 9-9 The structure of fluoride ion selective electrode

$$E = K - \frac{2.303RT}{F}\lg a_{F^-} \tag{9-24}$$

A galvanic cell is composed of F^- selective electrode as an indicator electrode and SCE as a reference electrode. It can be written as:

(−) Ag | AgCl(s) | NaCl(0.1mol/L), NaF(0.1mol/L) | LaF_3 | $F^-(c_x)$ | KCl(sat'd), Hg_2Cl_2 | Hg(+)

If the LaF_3 crystal film is replaced with AgCl, AgBr, AgI, CuS, PBS or Ag_2S, the ion selective electrode can be made for the determination of Ag^+, Cu^{2+}, Pb^{2+}, S^{2-}, etc.

Ammonia sensitive electrode (氨气敏电极, NH_3) is composed of glass pH electrode as primary electrode, Ag/AgCl as reference electrode, 0.1mol/L NH_4Cl solution as internal electrolyte, and polytetrafluoroethylene microporous film as permeable membrane (Figure 9-10) to detect the concentration of NH_4^+. Before measurement, a certain amount of NaOH solution is added into the sample solution to change NH_4^+ into NH_3 that cross the breathable membrane into 0.1mol/L NH_4Cl solution and cause pH change.

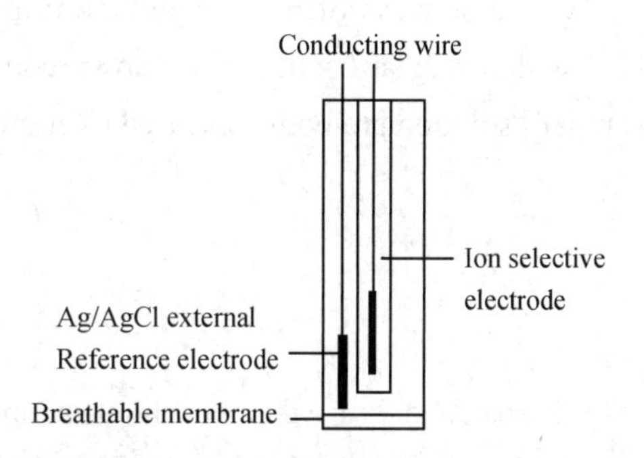

Figure 9-10 The structure of gas sensitive electrode

Enzyme electrode (酶电极) is made by modifying the primary electrode with a biological enzyme membrane or enzyme substrate membrane which can react with the analyte. If the urease is fixed on the ammonium ion electrode, it becomes the urease electrode (尿素酶电极). Urea undergoes the following reactions under the catalysis of urease.

$$NH_2CONH_2 + 2H_2O \xrightarrow{\text{Urease}} 2NH_4^+ + CO_3^{2-}$$

The urea content is determined indirectly by the potential response produced by the ammonium ion electrode.

2.3 Conditions and methods of quantitative analysis

2.3.1 Conditions of quantitative analysis

The ion selective electrode as the indicator electrode and the saturated calomel electrode as the reference electrode, are inserted into the sample solution to form the galvanic cell. Then ion activity or concentration is measured via the cell potential.

$$E = E_{SCE} - E_{ISE} = E_{SCE} - (K' \pm S\lg c) = K'' \mp S\lg c \tag{9-25}$$

2.3.2 Methods of quantitative analysis

(1) Standard contrast method

In standard contrast method, the potential of standard solution and sample solution are both measured. According to the Equation 9-25, the concentration of the analyte can be calculated as followed:

$$\lg c_x = \lg c_s \mp \frac{E_s - E_x}{S} \tag{9-26}$$

(2) Standard curve method

In standard curve method, a series of standard solution with different concentration are prepared and their potentials are gradually detected. The plot of lg$c$ with the measured E is usually a straight line within a certain concentration range, called the standard curve (Figure 9-11). Then, E_x of the sample solution is measured and the concentration of the ion c_x is quantificationally determined through the standard curve. The standard curve method requires that the standard solution and the sample solution have similar composition and ionic strength, so it is suitable for simple sample system.

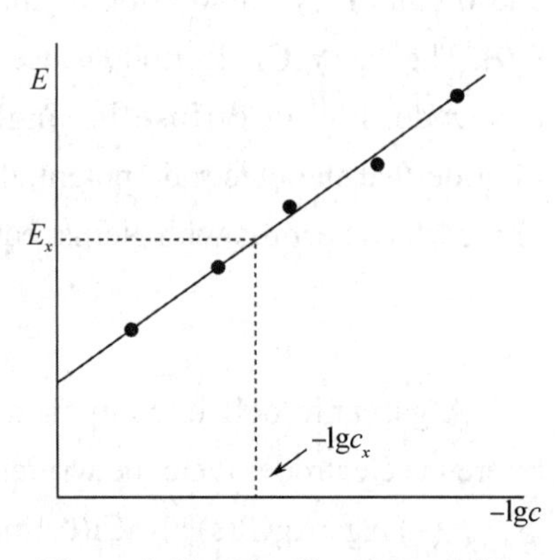

Figure 9-11 The illustration of the standard curve

(3) Standard addition method

Two same parts of the sample solution taken and one part is added with a standard solution of known concentration. The potential of sample solution with or without a standard solution are both measured (Requirement: $V_s < V_x/100$, $c_s > 100\ c_x$).

$$E_1 = K' \mp \frac{2.303}{nF} \lg c_x$$

$$E_2 = K'' \mp \frac{2.303}{nF} \lg \frac{c_x V_x + c_s V_s}{V_x + V_s}$$

Where c_x and V_x are the concentration and volume of sample solution; c_s and V_s are the concentration and volume of a standard solution; E_1 and E_2 are the potential of sample solution without and with the addition of standard solution, respectively.

Because $V_s < V_x/100$, so K' and K'' are similar.

$$10^{\frac{\Delta E}{S}} = \frac{c_x V_x + c_s V_s}{(V_x + V_s) c_x}$$

$$c_x = \frac{c_s V_s}{(V_x + V_s) 10^{\Delta E/S} - V_x} \tag{9-27}$$

Where V_x, c_s and V_s are all known and S can be obtained via the experiments, thus c_x can be calculated.

In standard contrast method and the standard curve method, the same amount of total ionic strength adjustment buffer (总离子强度调节缓冲剂, TISAB) should be added to the standard solution and the sample solution to make the activity coefficient and ionic strength of these two solutions consistent. The standard addition method does not need to add TISAB, and does not need to draw standard curve, so it is easy to operate and suitable for complicated system analysis.

3. Measurement error of direct potentiometry

Due to the influence of many factors such as electrode stability, liquid junction potential, and temperature fluctuation, there is more than ± 1mV error in the measurement of EMF by direct potentiometry. The relative error of sample concentration $\left(\frac{\Delta c}{c}\%\right)$ caused by the measurement error of cell potential difference (ΔE) can be obtained by differential Equation 9-29

$$\Delta E = \frac{RT}{nF} \times \frac{\Delta c}{c} \tag{9-28}$$

$$\frac{\Delta c}{c}\% = \frac{nF\Delta E}{TR} \times 100 \approx 3900 \times n\Delta E \quad (T{=}25°\text{C}) \tag{9-29}$$

4. Introduction to electrochemical biosensor and microelectrode technology

Electrochemical methods have attracted increasing attention due to their advantages of simple operation, high accuracy and selectivity, and satisfactory sensitivity. All kinds of electrochemical methods have been employed for microscopic, intravital, fast and automatic analysis in biology, chemistry, physics, and medicine fields. With the development of science and technology, a variety of new electrodes and sensors are rapidly emerging.

4.1 Electrochemical biosensor

A biosensor using an electrochemical electrode as a signal converter is called an electrochemical biosensor (电化学生物传感器). It consists of a signal converter and a sensor, in which the signal converter mainly include an electrochemical electrode and an ion-sensitive field effect transistor (离子敏场效应晶体管，ISFET). Enzyme sensors (酶传感器) are a kind of biosensors via enzyme catalysis reaction, such as glucose sensor, cholesterol sensor, lactic acid sensor, hydrogen peroxide sensor, phenylalanine sensor, adenosine sensor, uric acid and urea sensor. Microbial sensors (微生物传感器), using microbial bacteria as signal converter, can realize real-time analysis of microbial metabolism in vivo, and it has many advantages, such as low cost and long activity. Immunosensors (免疫传感器), based on the specific interaction between antigens and antibodies, conduct qualitative and quantitative analysis according to the changes of electrical parameters in the reaction process, such as chorionic gonadotropin sensor, alpha-fetoprotein sensor, etc. Field-effect biosensors (场效应生物传感器) can be made up by combining enzymes or other biometric molecules with ion-selective FET.

4.2 Microelectrode

Microelectrodes is made up of platinum wire, carbon fiber or sensitive membrane. According to the different materials of the electrode, the microelectrodes are divided into lead, gold, silver, tungsten and carbon fiber electrode. According to the shape of the electrode, it can be divided into disk electrode, circular electrode, cylindrical electrode, spherical electrode and complex microelectrode.

Compared to the conventional electrodes, microelectrodes own many novel electrochemical properties. Microelectrodes can set up steady state quickly; the capacitance of the double electrode layer on the surface of microelectrodes is very low; it has small solution impedance. Thus it greatly improves the response speed of electrode and the detection sensitivity. The sensitivity of electrochemical analysis can be further improved

by signal amplification of combined microelectrodes and modified microelectrodes. These above factors are beneficial for the study of thermodynamics and kinetics of electrochemical reaction process. Moreover, microelectrodes require fewer samples, greatly reducing the consumption of toxic reagents, environmental pollution, and test cost. Usually, the radius of microelectrodes is less than 50μm, even up to the nanoscale, which can apply to a single cell for non-destructive analysis in bioassay. It is very valuable for life science research. For example, using microelectrodes to study brain, it can quantitatively determine the content of bioactive amine, and analyze the dopamine and 5-hydroxytyramine in a single nerve cell. The application of microelectrodes in electrochemical analysis has developed rapidly in recent years.

PPT

Section 5 Potentiometric Titration

Potentiometric titration (电位滴定法) is a titration analysis method which uses the change of electrode potential of indicator electrode to determine the titration end point. It can be used in acid-base, precipitation, coordination, oxidation-reduction and non-aqueous titration, especially in the case that it is difficult to determine the end point of titration in colored or turbid solution or to select the appropriate indicator. Compared with the chemical indicator method, it has the advantages of strong objectivity, high accuracy, no limitation of solution color and turbidity, and is easy to realize titration analysis automation. Potentiometric titrations provide data that are more reliable than data from titrations that use chemical indicators and are particularly useful with colored or turbid solutions and for detecting the presence of unsuspected species.

Potentiometric titrations offer additional advantages over direct potentiometry. Because the measurement is based on the titrant volume that causes a rapid change in potential near the equivalence point, potentiometric titrations are not dependent on measuring absolute values of E_{cell}. This characteristic makes the titration relatively free from junction potential uncertainties because the junction potential remains approximately constant during the titration.

1. Basic principles and devices

Figure 9-12 illustrates a typical apparatus for performing a manual potentiometric titration. The operator measures and records the cell potential (in units of millivolts or pH, as appropriate) after each addition of reagent. The titrant is added in large increments early in the titration and in smaller and smaller increments as the end point is approached (as indicated by larger changes in cell potential per unit volume).

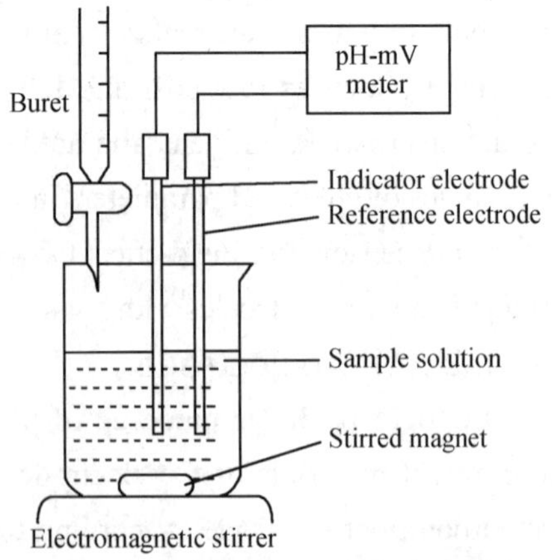

Figure 9-12 The instrument diagram of potentiometric titration

2. Endpoint determination method

In potentiometric titration, the electrode potential of the cell is measured every time after the titrant is added until exceeding the stoichiometric point. Generally, at the place far away from the stoichiometric

point, the volume of titrant drops is slightly larger; near the metering point, the volume of titrant should be added slowly. It is better to record the data once for each small part (0.10–0.05ml) added, and keep the volume of titrant added equal each time. Take the potentiometric titration of 2.433mmol NaCl with 0.1000mol/L $AgNO_3$ as an example. Electrode potential is recorded after 0.1000mol/L $AgNO_3$ is added into NaCl solution in Table 9-3.

Table 9-3 Potentiometric titration data for 2.433mmol NaCl with 0.1000mol/L $AgNO_3$

1 V (ml)	2 E (V)	3 ΔE (V)	4 ΔV (ml)	5 $\Delta E/\Delta V$ (V/ml)	6 $\overline{V}$ (ml)	7 $\Delta(\Delta E/\Delta V)$ (V/ml)	8 $\Delta\overline{V}$ (ml)	9 $\Delta^2E/\Delta V^2$ (V/ml^2)
22.00	0.123	0.015	1.00	0.015	22.50			
23.00	0.138	0.036	1.00	0.036	23.50	0.021	1.00	0.021
24.00	0.174	0.009	0.10	0.09	24.05	0.054	0.55	0.098
24.10	0.183	0.011	0.10	0.11	24.15	0.02	0.10	0.2
24.20	0.194	0.039	0.10	0.39	24.25	0.28	0.10	2.8
24.30	0.233	0.083	0.10	0.83	24.35	0.44	0.10	4.4
24.40	0.316	0.024	0.10	0.24	24.45	−0.59	0.10	−5.9
24.50	0.340	0.011	0.10	0.11	24.55	−0.13	0.10	−1.3
24.60	0.310	0.024	0.40	0.06	24.80	−0.05	0.25	−0.2
25.00	0.375							

2.1 Graphical method

2.1.1 Titration curve method (E~V curve)

The curve is plotted with electrode potential (E) as the longitudinal ordinate and titrant volume (V) as the horizontal ordinate, as shown in Figure 9-13A.The horizontal ordinate value corresponding to the turning point of the curve is end point (V_{ep}) of the titration. This method is simple to use, but it requires obvious titration break. If the titration break is not obvious, it should be determined by first-derivative curve method ($\Delta E/\Delta V \sim V$) or second-derivative curve method ($\Delta^2E/\Delta V^2 \sim V$).

2.1.2 First-derivative curve method ($\Delta E/\Delta V$–V)

The curve is plotted with the first derivative ($\Delta E/\Delta V$) as the longitudinal ordinate and titrant volume (V) as the horizontal ordinate, as shown in Figure 9-13B. The peak of the curve indicates the maximal change in potential per unit volume of titrant, corresponding to titrant volume at the end point (V_{ep}). If the titration curve is symmetrical, the point of maximum slope coincides with the equivalence point. For the asymmetrical titration curves that are observed when the titrant and analyte half-reactions involve different numbers of electrons. A smaller titration error occurs if the point of maximum slope is used, compared to titration curve method.

2.1.3 Second-derivative curve method ($\Delta^2E/\Delta V^2 \sim V$)

The curve is plotted with the first derivative ($\Delta^2E/\Delta V^2$) as the longitudinal ordinate and titrant volume (V) as the horizontal ordinate, as shown in Figure 9-13C. This method is used to analyze the analytical signal in some automatic titrators. The point at which the second derivative crosses zero is the inflection point (V_{ep}), which is taken as the end point of the titration, and this point can be located quite precisely.

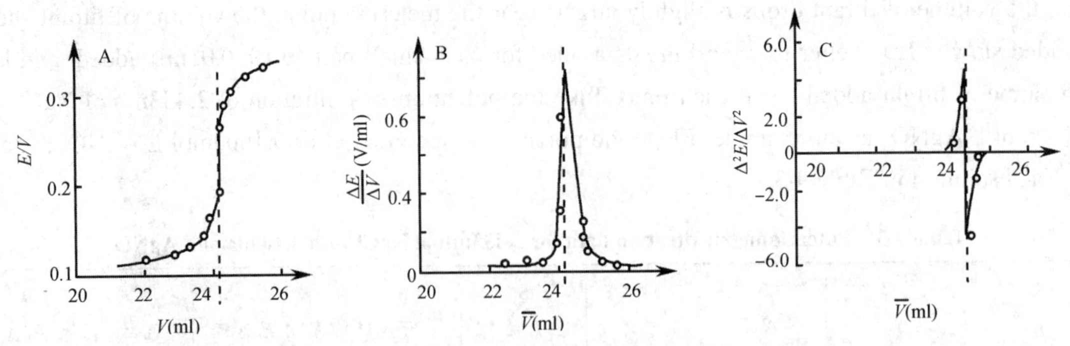

Figure 9-13 Determination of the end point in potentiometric titration of 2.433mmol of chloride ion with 0.1000mol/L silver nitrate

(A) Titration curve; (B) First-derivative curve; (C) Second-derivative curve

2.2 Second order differential quotient interpolation

As graphical method is rather complicated, a series of data near the end point are needed to draw curves. Since the second derivative curve near the measuring point is approximate to a straight line, the end point (V_{ep}) can be calculated by second order differential quotient interpolation.

$$\frac{A-0}{V_A-V_{ep}}=\frac{A-B}{V_A-V_B}$$

$$V_{ep}=\frac{A}{A-B}\times(V_B-V_A)+V_A-V_B$$

When V_{AgNO_3}=24.30ml, $\Delta^2E/\Delta V^2$=4.4; when V_{AgNO_3}=24.40ml, $\Delta^2E/\Delta V^2$=5.9, therefore

$$V_{ep}=\frac{4.4}{4.4+5.9}\times(24.40-24.30)+24.30=24.34\text{ml}$$

3. Application examples

Provided a suitable indicator electrode, all kinds of titration analysis can be used in potentiometric titration, such as acid-base titration, precipitation titration, coordination titration, oxidation-reduction titration as Table 9-4 illustrated.

Table 9-4 Electrode used in different titration method

Method	Electrode system	Instructions
Acid-base titration	pH glass electrode/SCE pH combination electrode	pH glass electrode should be stored in pure water and the pH combination electrode should be stored in 3mol/L KCl solution
Nonaqueous titration	pH glass electrode/SCE	SCE is filled with saturated KCl/anhydrous methanol solution to avoid the interference of water exudation, or double salt bridge SCE is used. pH glass electrode is the same as in acid-base titration
Precipitation titration	Ag/SCE	Double salt bridge SCE should be used to avoid interference from Cl^- of SCE
	Ag/pH glass electrode	pH glass electrode is used as a reference electrode. A few HNO_3 can be added into sample solution to keep the potential of pH glass electrode stable
	ISE/SCE	The same as Ag/SCE

continued

Method	Electrode system	Instructions
Coordination titration	pM Hg/SCE	Add 3–5 drops of 0.05mol/L HgY^{2-} solution to the sample solution in advance. It is suitable for the metal ions whose K_{MY} is less than $K_{HgY^{2-}}$. Used under suitable pH range
Redox titration	Pt/SCE	The platinum electrode should be immersed in HNO_3 solution containing a small amount of $FeCl_3$ or chromic acid to clean the electrode surface

Section 6 Biamperometric Titration

Biamperometric titration is also called dead-stop titration (永停滴定法). Two same metal electrodes (such as Pt) with a low voltage (100–300mV) between them are inserted into the sample solution. The titration end point is determined according to the sudden change of electrolytic current during titration.

1. Basic principles and devices

As shown in Figure 9-14, the biamperometric titration device has three main parts: two Pt electrodes in the sample solution to form electrolytic cell; a power circuit with additional small voltage; a sensitive galvanometer for measuring the electrolytic current, etc. When two same Pt electrodes are inserted into the sample solution containing oxidation and reduction states of metals, there is no current and the potential between them is zero. If a small voltage is exerted at both ends, the biamperometric titration device becomes an electrolytic cell. Take Fe^{3+}/Fe^{2+} pair as an example, the electrolytic reaction is as follow:

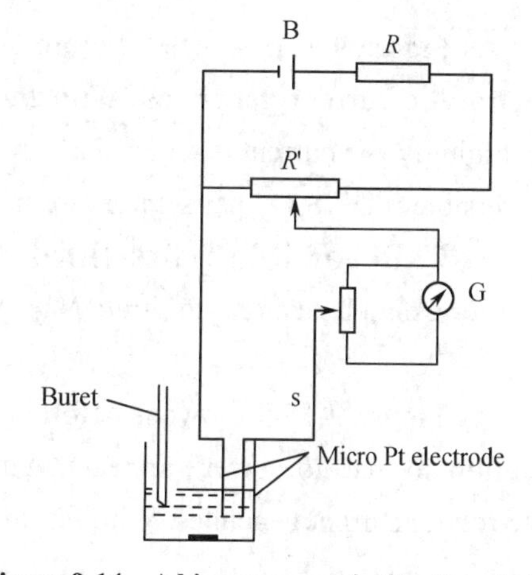

Figure 9-14 A biamperometric titration device

Anode (Positive electrode), oxidation reaction: $Fe^{2+} \rightleftharpoons Fe^{3+}+e$

Cathode (Negative electrode), reduction reaction: $Fe^{3+}+e \rightleftharpoons Fe^{2+}$

Redox reaction of Fe^{3+}/Fe^{2+} induces the generation of the electrolytic current. As well as Fe^{3+}/Fe^{2+}, some other pairs, such as Ce^{4+}/Fe^{2+}, I_2/I^-, Br_2/Br^-, and HNO_2/NO, can form an electrolytic cell with two Pt electrodes. These pair with the property that the electrolytic current generates after addition with a small voltage, is called reversible pairs (可逆电对). In contrary to reversible pairs, irreversible pairs (不可逆电对) cannot provide the electrolytic current after addition with a small voltage, such as $S_4O_6^{2-}/S_2O_3^{2-}$. Although $S_2O_3^{2-}$ loses 2e to be $S_4O_6^{2-}$ on anode electrode, $S_4O_6^{2-}$ cannot obtain 2e to be $S_2O_3^{2-}$ on cathode electrode. Therefore, there is no current in electric circuit.

Biamperometric titration method is based on the relationship between the electrolytic current and titrant volume to determine the titration endpoint. The electrolytic current is measured every time after the titrant is added. I–V curve is plotted with the current as the longitudinal ordinate (纵坐标) and titrant volume (V) as the horizontal ordinate (横坐标).

2. Endpoint determination method

According to the properties of the electric pairs participating in the reaction, there are generally three types of titration.

2.1 Reversible pairs titrate reversible pairs

For example, using Ce^{4+} to titrate Fe^{2+}, the reaction is as follow:

$$Ce^{4+} + Fe^{2+} \rightleftharpoons Ce^{3+} + Fe^{3+}$$

Before the titration, there is only Fe^{2+} in solution and cannot form Fe^{3+}/Fe^{2+} pair so that no electrolytic current appears. With the addition of Ce^{4+} to react with Fe^{2+}, the concentration of Fe^{3+} increases to provide Fe^{3+}/Fe^{2+} pair, therefore the electrolytic current also increases. As Figure 9-15A showed, when $c_{Fe^{3+}} = c_{Fe^{2+}}$, the current reaches the maximum value. Then, the current gradually decreases due to the concentration decrease of Fe^{2+} if Ce^{4+} is continually added and reach the minimum value at endpoint. After endpoint, the electrolytic current increase with the existence of Ce^{4+}/Ce^{3+} pair.

2.2 Irreversible pairs titrate reversible pairs

For example, using $Na_2S_2O_4$ to titrate I_2 containing excess KI, the reaction is as follow:

$$I_2 + 2S_2O_3^{2-} \rightleftharpoons S_4O_6^{2-} + 2I^-$$

As Figure 9-15B showed, before stoichiometric point, reversible pair I_2/I^- is present so that the electrolytic current generates. With the addition of $Na_2S_2O_4$, the concentration of I_2 is decreased. Accordingly, the current is reduced and reaches to zero at stoichiometric point. After stoichiometric point, irreversible $S_4O_6^{2-}/S_2O_3^{2-}$ pairs and I^- are in solution, so the current cannon produce current.

2.3 Reversible pairs titrate irreversible pairs

For example, using I_2 to titrate $Na_2S_2O_4$, the chemical reaction is as follow:

$$I_2 + 2S_2O_3^{2-} \rightleftharpoons S_4O_6^{2-} + 2I^-$$

As Figure 9-15C showed, before stoichiometric point, there are irreversible $S_4O_6^{2-}/S_2O_3^{2-}$ pairs in solution so that no current appears. After stoichiometric point, reversible I_2/I^- pairs forms, then the electrolytic current generates. With the addition of I_2, the electrolytic current also raises.

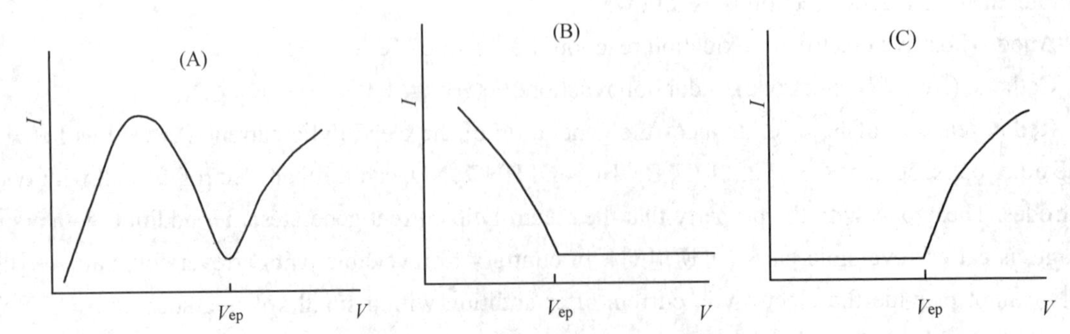

Figure 9-15 (A) Titration curve of Fe^{2+} with Ce^{4+}; (B) Titration curve of I_2 with $S_2O_3^{2-}$; (C) Titration curve of $S_2O_3^{2-}$ with I_2

3. Application examples

Biamperometric titration method has many advantages of simple instrument, fast and convenient operation, direct and accurate data, and being easy to realize automatic titration, etc. In Chinese Pharmacopoeia, biamperometric titration is applied to diazotization titration and *Karl Fischer* water determination.

(1) Endpoint determination in measuring water by *Karl Fischer* method

The reaction between H_2O and *Karl Fischer* titrant is as follow:

$$I_2 + SO_2 + 3\,C_5H_5N + CH_3OH + H_2O \longrightarrow 2\,C_5H_5N^{+}H\,I^{-} + C_5H_5N^{+}H\,SO_4CH_3^{-}$$

Before the addition of *Karl Fischer* regent, there is no reversible pairs in solution that galvanometer remains zero. After stoichiometric point, reversible I_2/I^- pairs form to produce electrolytic current and galvanometer pointer position offset. The end point of titration is the point at which the pointer just shifts.

(2) Endpoint determination using sodium nitrite

Sodium nitrite is often used to titrate aromatic amines in acidic solution. Their reaction is present as follow:

$$R-C_6H_4-NH_2 + NaNO_2 + 2HCl \rightleftharpoons [R-C_6H_4-N^{+}]Cl^{-} + 2H_2O + NaCl$$

Before the addition of sodium nitrite, there is irreversible pairs in solution and galvanometer remains zero. After stoichiometric point, with a part of excess HNO_2 decomposing to NO, reversible HNO_2/NO pair forms and electrolytic current is produced. The end point of titration is the point at which the pointer just shifts.

Anode: $NO+H_2O \rightleftharpoons HNO_2+H^++e$

Cathode: $HNO_2+H^++e \rightleftharpoons NO+H_2O$

重点小结

电化学分析方法分为电位分析法、电解分析法、伏安法和电导分析法等，其中电位分析法可分为直接电位法和电位滴定法，电流滴定法包括单指示电极电流滴定法和双指示电极电流滴定法，后者又称永停滴定法。本章节主要介绍电位分析法基本原理，利用直接电位法、电位滴定法和双指示电极电流滴定法进行滴定分析。

电位分析法的基本原理基于原电池中物质在电极上的氧化还原反应。需由两支电极和待测溶液构成原电池，其中一支电极电位随电解质溶液中待测离子活度（或浓度）变化而改变的电极称为指示电极（indicator electrode），另一支电极的电极电位与待测离子活度（或浓度）无关，其电极电位值保持恒定，仅提供测量参考作用，称为参比电极（reference electrode）。常用指示电极电极通常有金属基电极和离子选择电极，参比电极有标准氢电极、饱和甘汞电极、双盐桥饱和甘汞电极和银 - 氯化银电极。

直接电位法根据待测组分的电化学性质，选择合适的指示电极和参比电极，插入试液中组成原电池，测量原电池的电动势，根据 Nernst 方程式给定的电极电位与待测组分活度的关系，求出

待测组分含量的方法。主要用于测量溶液的 pH 值和溶液中其他离子的活（浓）度。pH 值的测定所用电极为 pH 玻璃电极。溶液中其他离子的活（浓）度的测定使用电极如氟离子选择电极，氨敏电极和酶电极等。定量分析的方法包括标准对照法、标准曲线法和标准加入法。

电位滴定法是利用滴定过程中指示电极的电极电位（或为电池电动势）的变化来确定滴定终点的滴定分析法。终点确定方法包括 E–V 曲线法、$\triangle E/\triangle V$–V 曲线法、$\triangle^2 E/\triangle V^2$–$V$ 曲线法和二阶微商 - 线性内插法。

双指示电极电流滴定法：又称永停滴定法，根据电流变化来确定终点，电流的三种变化曲线和重点的确定。

题库

目 标 检 测

1. Galvanic cell and electrolytic cell.
2. Reference electrode and indicator electrode.
3. Liquid junction potential.
4. Asymmetry potential.
5. Reversible pairs and irreversible pairs.
6. Potentiometric titration.
7. Biamperometric titration.
8. What are the main differences between direct potential method and potentiometric titration?

9. Try to describe the concept, principle, characteristics of potentiometric titration and the method to determine the end point.

10. Compare these several common indicator electrodes (from classification, expression, electrode reaction, electrode potential, phase interface.)

11. Calculate the potential of the following cell at 25°C and indicate whether the Ag electrode shown would act as an anode or a cathode.

$$Cu \mid Cu^{2+}\ (0.0100mol/L) \parallel Cl^-\ (0.0100mol/L) \mid AgCl\ (s) \mid Ag$$

12. The internal reference electrode of fluoride ion selective electrode is Ag/AgCl electrode and $E^{\ominus}_{AgCl/Ag}$=0.2223V. The internal reference solution is composed of 0.10mol/L NaCl and 1.0×10^{-3}mol/L NaF, Calculate the potential of this fluoride ion selective electrode when it is immersed in 1.0×10^{-5}mol/L F^- with pH=7 at 25°C. $K^{pot}_{F^-, OH^-}$=0.10)

13. The pH glass electrode and the saturated calomel electrode (positive electrode) were inserted into the standard buffer solution with a pH of 6.86, and the measured potential E (or EMF) was 0.345V. When the standard buffer solution was replaced by the unknown solution, the measured potential E (or EMF) is 0.298V. Calculate the pH of the unknown solution at 25°C.

Tip: Adopt the formula of measuring pH value of solution by "two-step method".

14. A Ca electrode is immersed in 25.00ml sample solution to detect the concentration of Ca^{2+} and a reference electrode is used as positive electrode. The potential of this electrochemical cell is 0.4695V at 25°C. After addition of 1.00ml standard $CaCl_2$ (5.45×10^{-2}mol/L) solution, the potential decreases to 0.4117V. Calculate the concentration of Ca^{2+} in the sample solution.

15. The following is a partial data sheet for titration of 50.00ml of a weak acid with a standard solution of 0.1250mol/L NaOH.

Volume (ml)	0.00	0.80	20.00	36.00	39.20	39.72	40.00	40.08	40.80
pH	2.40	3.21	3.81	4.76	5.56	6.51	8.25	10.00	11.00

(1) After processing these data by $\Delta^2\text{pH}/\Delta V^2-V$ curve method, please calculate the volume of titration end-point and the concentration of acid solution.

(2) Please calculate the dissociation constant of the weak acid solution.

Tip: ① The final volume of titration end-point can be obtained by mathematical interpolation method. ② At the hemi-neutralization point, the pH of the solution is equal to $\text{p}K_a$.

Answer to Exercises

第一章

略

第二章

5. 0.028ml
7. 33.45, 0.025, 0.075%, 0.037, 0.11%; 0.1043, 0.0002, 0.20%, 0.0003, 0.28%
8. 5, 4, 3
9. 1.237, 1.238, 0.135, 2.1, 2.00
10. 36.90, 3.50, 40.0, 1.21×10^{-5}, 5.4911
13. $\mu = 16.883 \pm 0.066$
14. no significant difference
15. no significant difference
16. should be retained
17. should be retained

第三章

8. 53.44% w/w Fe_3O_4
9. 98.68% w/w KBr

第四章

4. 0.1115mol/L
5. 12mol/L, 8.3ml
6. 0.1001mol/L
7. 0.2097mol/L
8. 26.67%
9. 0.1146mol/L
10. 0.2587mol/L

第五章

1. (1) H_2CO_3, CH_3COOH, H_3O^+, $C_6H_5NH_3^+$, NH_4^+, HAc, HS^-;
   (2) NO_3^-, OH^-, HPO_4^{2-}, CO_3^{2-}, $C_2O_4^{2-}$, HS^-, PO_4^{3-}
4. 8.88, 5.12
5. $pH=pK_{In}\pm1$

11. (1) Yes, pH=8.22, phenol red or phenolphthalein
    (2) No
    (3) Yes, pH=8.45, phenolphthalein or thyme blue
    (4) No
    (5) No
    (6) Yes, pH=5.64, methyl red
12. (1) Yes, K_{a1} and K_{a2} of H_3PO_4 can be titrated separately.
    (2) No, 0.10 mol/L $H_2C_2O_4$ cannot be titrated separately
    (3) Yes.
    (4) Yes.
14. (3) liquid ammonia
18. 4.70, 4.76
19. 3.47
20. 7.00, 9.55; 2.00, 12.00, 2.00, 17.10
21. 337.1, 1.26×10^{-5}, 8.76
22. 73.97%
23. 86.0%
24. 1.500mmol, 2.000mmol
25. 60.10%
26. NaOH%, 59.73%; Na_2CO_3%, 39.57%
27. 0.2%, 0.02%
28. −0.56%, 0.2%
29. 0.1388mol/L, 0.1500mol/L
30. 0.1079mol/L
31. 7.696%
32. 99.5%

第六章

2. (1) positive; (2) none; (3) positive; (4) negative.
3. (1) negative; (2) none; (3) negative; (4) none.
4. (1) negative; (2) none; (3) positive; (4) negative.
5. titrant: $AgNO_3$, indicator: 1–2 ml 5% K_2CrO_4, pH: 6.5–10.5, vigorous shaking.
6. titrant: $AgNO_3$, KSCN, indicator: $NH_4Fe(SO_4)_2 \cdot 12H_2O$,
   pH: strongly acidic solution, filtrate AgCl or add nitrobenzene to protect AgCl
7. titrant: $AgNO_3$, indicator: fluorescein, pH: 7-10, add dextrin before titrating,
   avoid bright lights during the titration.
8. (1) Volhard; (2) Volhard; (3) Volhard; (4) Mohr / Volhard;
   (5) Volhard; (6) Volhard / Fajans; (7) Mohr / Volhard / Fajans; (8) Volhard / Fajans.
9. 89.36%
10. KCl% = 5.919, KBr% = 46.53
11. 53.72%
12. 97.06%, qualified.

第七章

6. 0.00918mol/L
7. (a) 32.28ml (b) 14.98ml (c) 32.28ml
8. (a) 34.84ml (c) 45.99ml (e) 32.34ml
9. 3.244%
10. 1.216%
12. 1.4×10^{-3} mol/L
13. 2.0×10^{-5} mol/L
14. $[Ag^+] = 3.4\times10^{-16}$mol/L; $[Ag(S_2O_3)^-] = 2.2\times10^{-7}$mol/L; $[Ag(S_2O_3)_2^{3-}]=9.9\times10^{-3}$mol/L
15. 184.0mg/L Fe^{3+} and 213.1mg/L Fe^{2+}
17. 55.16% Pb and 44.86% Cd
19. (a) 2.78×10^7; (b) 3.90×10^{17}
20. (a) (1) 1.60 (2) 2.12 (3) 4.80 (4) 7.22; (b) (1) 1.60 (2) 2.12(3) 9.87 (4) 17.37
21. 1.26
22. 2.10×10^{14}
23. 83.75% ZnO and 0.230% Fe_2O_3
24. 13.72% Cr, 56.82% Ni, and 27.44% Fe
25. (a) 4.73×10^9 (b) 1.13×10^{12} (c) 7.53×10^{13}
26. (a)2.7×10^{-10} (b) 0.57
27. (a)2.5×10^7 (b) 4.5×10^{-5}mol/L
28. 1.70, 2.18, 2.81, 3.87, 4.87, 6.85, 8.82, 10.51, 10.82
29. ∞, 10.30, 9.52, 8.44, 7.43, 6.15, 4.88, 3.20, 2.93
30. 4.6×10^{-11} mol/L
31. (a) 11.08 (b) 11.09 (c) 12.35 (d) 15.03 (e) 17.69
32. 1. with metal ion indicators; 2. with a mercury electrode; 3. with an ion-selective electrode; 4. with a glass electrode
33. (wine-red, blue)
34. Buffer (a): yellow→blue;other buffers: violet→blue, which is harder to see
35. (a) 570.5mg/L (b) 350.5mg/L (c) 185.3mg/L
36. $[Ca^{2+}] = 0.0017$mol/L, $[Ca(NO_3)^+] = 0.0033$mol/L

第八章

4. 0.12 V
5. 8.7×10^7
11. (a) 1.433×10^{-3}mol
    (b) 0.07632mol/L
    (c) 12.83%
17. 0.0992g/ml
18. 78.02%
19. 94.28%

第九章

11. 0.062V, Ag electrode act as an anode.
12. 0.399V
13. 6.06
14. 2.33×10^{-5}mol/L
15. (1) 40.00mL, (2) 1.55×10^{-4}

Appendix

Appendix 1 Atomic Weight

(Based on Carbon-12)

Number	Element	Symbol	Atomic weight	Number	Element	Symbol	Atomic weight
1	Hydrogen	H	1.00794(7)	26	Iron	Fe	55.845(2)
2	Helium	He	4.002602(2)	27	Cobalt	Co	58.933195(5)
3	Lithium	Li	6.941(2)	28	Nickel	Ni	58.6934(4)
4	Beryllium	Be	9.012182(3)	29	Copper	Cu	63.546(3)
5	Boron	B	10.811(7)	30	Zinc	Zn	65.38(2)
6	Carbon	C	12.0107(8)	31	Gallium	Ga	69.723(1)
7	Nitrogen	N	14.0067(2)	32	Germanium	Ge	72.64(1)
8	Oxygen	O	15.9994(3)	33	Arsenic	As	74.92160(2)
9	Fluorine	F	18.9984032(5)	34	Selenium	Se	78.96(3)
10	Neon	Ne	20.1797(6)	35	Bromine	Br	79.904(1)
11	Sodium	Na	22.98976928(2)	36	Krypton	Kr	83.798(2)
12	Magnesium	Mg	24.3050(6)	37	Rubidium	Rb	85.4678(3)
13	Aluminum	Al	26.9815386(8)	38	Strontium	Sr	87.62(1)
14	Silicon	Si	28.0855(3)	39	Yttrium	Y	88.90585(2)
15	Phosphorus	P	30.973762(2)	40	Zirconium	Zr	91.224(2)
16	Sulphur	S	32.065(5)	41	Niobium	Nb	92.90638(2)
17	Chlorine	Cl	35.453(2)	42	Molybdenum	Mo	95.96(2)
18	Argon	Ar	39.948(1)	43	Technetium	Tc	[98]
19	Potassium	K	39.0983(1)	44	Ruthenium	Ru	101.07(2)
20	Calcium	Ca	40.078(4)	45	Rhodium	Rh	102.90550(2)
21	Scandium	Sc	44.955912(6)	46	Palladium	Pd	106.42(1)
22	Titanium	Ti	47.867(1)	47	Silver	Ag	107.8682(2)
23	Vanadium	V	50.9415(1)	48	Cadmium	Cd	112.411(8)
24	Chromium	Cr	51.9961(6)	49	Indium	In	114.818(3)
25	Manganese	Mn	54.938045(5)	50	Tin	Sn	118.710(7)

continued

Number	Element	Symbol	Atomic weight	Number	Element	Symbol	Atomic weight
51	Antimony	Sb	121.760(1)	82	Lead	Pb	207.2(1)
52	Tellurium	Te	127.60(3)	83	Bismuth	Bi	208.98040(1)
53	Iodine	I	126.90447(3)	84	Polonium	Po	[209]
54	Xenon	Xe	131.293(6)	88	Radium	Ra	[226]
55	Caesium	Cs	132.9054519(2)	89	Actinium	Ac	[227]
56	Barium	Ba	137.327(7)	90	Thorium	Th	232.03806(2)
57	Lanthanum	La	138.90547(7)	91	Protactinium	Pa	231.03588(2)
58	Cerium	Ce	140.116(1)	92	Uranium	U	238.02891(3)
59	Praseodymium	Pr	140.90765(2)	93	Neptunium	Np	[237]
60	Neodymium	Nd	144.242(3)	94	Plutonium	Pu	[244]
61	Promethium	Pm	[145]	95	Americium	Am	[243]
62	Samarium	Sm	150.36(2)	96	Curium	Cm	[247]
63	Europium	Eu	151.964(1)	97	Berkelium	Bk	[247]
64	Gadolinium	Gd	157.25(3)	98	Californium	Cf	[251]
65	Terbium	Tb	158.92535(2)	99	Einsteinium	Es	[252]
66	Dysprosium	Dy	162.500(1)	100	Fermium	Fm	[257]
67	Holmium	Ho	164.93032(2)	101	Mendelevium	Md	[258]
68	Erbium	Er	167.259(3)	105	Dubnium	Db	[268]
69	Thulium	Tm	168.93421(2)	106	Seaborgium	Sg	[271]
70	Ytterbium	Yb	173.054(5)	107	Bohrium	Bh	[272]
71	Lutetium	Lu	174.9668(1)	108	Hassium	Hs	[277]
72	Hafnium	Hf	178.49(2)	109	Meitnerium	Mt	[276]
73	Tantalum	Ta	180.94788(2)	110	Darmstadtium	Ds	[281]
74	Tungsten	W	183.84(1)	111	Roentgenium	Rg	[280]
75	Rhenium	Re	186.207(1)	112	Ununbium	Uub	[285]
76	Osmium	Os	190.23(3)	113	Ununtrium	Uut	[284]
77	Iridium	Ir	192.217(3)	114	Ununquadium	Uuq	[289]
78	Platinum	Pt	195.084(9)	115	Ununpentium	Uup	[288]
79	Gold	Au	196.966569(4)	116	Ununhexium	Uuh	[293]
80	Mercury	Hg	200.59(2)	118	Ununoctium	Uuo	[294]
81	Thallium	Tl	204.3833(2)				

Appendix 2 Formula Weights

Compound	Formula weight	Compound	Formula weight
AgBr	187.77	KH_2PO_4	136.09
AgCl	143.32	$KHSO_4$	136.17
AgI	234.77	KI	166
$AgNO_3$	169.87	KIO_3	214
Al_2O_3	101.96	$KIO_3 \cdot HIO_3$	389.91
As_2O_3	197.84	$KMnO_4$	158.03
$BaCl_2 \cdot 2H_2O$	244.26	KNO_2	85.104
BaO	153.33	KOH	56.106
$Ba(OH)_2 \cdot 8H_2O$	315.47	K_2PtCl_6	486
$BaSO_4$	233.39	KSCN	97.181
$CaCO_3$	100.09	$MgCO_3$	84.314
CaO	56.077	$MgCl_2$	95.211
$Ca(OH)_2$	74.093	$MgSO_4 \cdot 7H_2O$	246.48
CH_3COOH	60.052	$MgNH_4PO_4 \cdot 6H_2O$	245.41
$C_6H_4COOHCOOK$	204.22	MgO	40.304
C_6H_5COONa	144.1	$Mg(OH)_2$	58.32
CO_2	44.01	$Mg_2P_2O_7$	222.55
CuO	79.545	$Na_2B_4O_7 \cdot 10H_2O$	381.37
Cu_2O	143.09	NaBr	102.89
$CuSO_4 \cdot 5H_2O$	249.69	NaCl	58.443
FeO	71.844	Na_2CO_3	105.99
Fe_2O_3	159.69	$NaHCO_3$	84.007
$FeSO_4 \cdot 7H_2O$	278.02	$Na_2HPO_4 \cdot 12H_2O$	358.14
$FeSO_4 \cdot (NH_4)_2SO_4 \cdot 6H_2O$	392.14	$Na_2H_2Y \cdot 2H_2O$	372.24
H_3BO_3	61.833	$NaNO_2$	68.995
HCl	36.461	Na_2O	61.979

continued

Compound	Formula weight	Compound	Formula weight
$HClO_4$	100.46	$NaOH$	39.997
$H_2C_2O_4 \cdot 2H_2O$	126.07	$Na_2S_2O_3$	158.11
HNO_3	63.013	$Na_2S_2O_3 \cdot 5H_2O$	248.18
H_2O	18.015	NH_3	17.031
H_2O_2	34.015	NH_4Cl	53.492
H_3PO_4	97.995	NH_4OH	35.046
H_2SO_4	98.079	$(NH_4)_3PO_4 \cdot 12MoO_3$	1876.6
I_2	253.81	$(NH_4)_2SO_4$	132.14
$KAl(SO_4)_2 \cdot 12H_2O$	474.39	$PbCrO_4$	323.19
KBr	119	PbO_2	239.2
$KBrO_3$	167	$PbSO_4$	303.26
KCl	74.551	P_2O_5	141.94
$KClO_4$	138.55	SiO_2	60.084
K_2CO_3	138.21	SO_2	64.064
K_2CrO_4	194.19	SO_3	80.063
$K_2Cr_2O_7$	242.19	ZnO	81.379

Appendix 3 Dissociation Constants for Acids (25℃)

Name	Formula	K_a	pK_a
Arsenic acid	H_3AsO_4	5.5×10^{-3}	2.26
		1.7×10^{-7}	6.76
		3.2×10^{-12}	11.29
Arsenious acid	H_2AsO_3	5.1×10^{-10}	9.29
Boric acid	H_3BO_3	5.4×10^{-10}	9.27(20℃)
Carbonic acid	H_2CO_3	4.5×10^{-7}	6.35
		4.7×10^{-11}	10.33
Chromic acid	H_2CrO_4	0.18	0.74
		3.2×10^{-7}	6.49
Hydrofluoric acid	HF	6.3×10^{-4}	3.20
Hydrocyanic acid	HCN	6.2×10^{-10}	9.21
Hydrogen sulfide	H_2S	8.9×10^{-8}	7.05
		1.0×10^{-19}	19.0
Hydrogen peroxide	H_2O_2	2.4×10^{-12}	11.62
Hypobromous acid	HBrO	2.8×10^{-9}	8.55
Hypochlorous acid	HClO	4.0×10^{-8}	7.40
Hypoiodous acid	HIO	3.2×10^{-11}	10.5
Iodic acid	HIO_3	0.17	0.78
Nitrous acid	HNO_2	5.6×10^{-4}	3.25
Perchloric acid	$HClO_4$		−1.6(20℃)
Periodic acid	HIO_4	2.3×10^{-2}	1.64
Phosphoric acid	H_3PO_4	6.9×10^{-3}	2.16
		6.2×10^{-8}	7.21
		4.8×10^{-13}	12.32
Phosphorous acid	H_3PO_3	5.0×10^{-2}	1.30(20℃)
		2.0×10^{-7}	6.70(20℃)

continued

Name	Formula	K_a	pK_a
Pyrophosphoric acid	$H_4P_2O_7$	0.12	0.91
		7.9×10^{-3}	2.10
		2.0×10^{-7}	6.70
		4.8×10^{-10}	9.32
Silicic acid	H_4SiO_4	1.6×10^{-10}	9.9(30℃)
		1.6×10^{-12}	11.8(30℃)
		1.0×10^{-12}	12.0(30℃)
		1.0×10^{-12}	12.0(30℃)
Sulfuric acid	H_2SO_4	1.0×10^{-2}	1.99
Sulfurous acid	H_2SO_3	1.4×10^{-2}	1.85
		6.3×10^{-8}	7.20
Water	H_2O	1.01×10^{-14}	13.995
Formic acid	HCOOH	1.8×10^{-4}	3.75
Acetic acid	CH_3COOH	1.7×10^{-5}	4.76
Acrylic acid	$H_2CCHCOOH$	5.6×10^{-5}	4.25
Benzoic acid	C_6H_5COOH	6.3×10^{-5}	4.20
Chloroacetic acid	$CH_2ClCOOH$	1.3×10^{-3}	2.87
Dichloroacetic acid	$CHCl_2COOH$	4.5×10^{-2}	1.35
Trichloroacetic acid	CCl_3COOH	0.22	0.66
Oxalic acid	$H_2C_2O_4$	5.6×10^{-2}	1.25
		1.5×10^{-4}	3.81
Adipic acid	$(CH_2CH_2COOH)_2$	3.9×10^{-5}	4.41(18℃)
		3.9×10^{-6}	5.41(18℃)
Malonic acid	$CH_2(COOH)_2$	1.4×10^{-3}	2.85
		2.0×10^{-6}	5.70
Succinic acid	$(CH_2COOH)_2$	6.2×10^{-5}	4.21
		2.3×10^{-6}	5.64
Maleic acid	$C_2H_2(COOH)_2$	1.2×10^{-2}	1.92
		5.9×10^{-7}	6.23
Fumaric acid	$C_2H_2(COOH)_2$	9.5×10^{-4}	3.02
		4.2×10^{-5}	4.38
Phthalic acid	$C_6H_4(COOH)_2$	1.1×10^{-3}	2.94
		3.7×10^{-6}	5.43

continued

Name	Formula	K_a	pK_a
meso-Tartaric acid	$(CHOHCOOH)_2$	6.8×10^{-4}	3.17
		1.2×10^{-5}	4.91
Salicylic acid	$C_6H_4OHCOOH$	1.0×10^{-3}	2.98(20℃)
2-Hydroxybenzoic acid		2.5×10^{-14}	13.6(20℃)
Malic acid	$HOCHCH_2(COOH)_2$	4.0×10^{-4}	3.4
		7.8×10^{-6}	5.11
Citric acid	$C_3H_4OH(COOH)_3$	7.4×10^{-4}	3.13
		1.7×10^{-5}	4.76
		4.0×10^{-7}	6.40
L-Ascorbic acid	$C_6H_8O_6$	9.1×10^{-5}	4.04
Phenol	C_6H_5OH	1.0×10^{-10}	9.99
Glycolic acid	$HOCH_2COOH$	1.5×10^{-4}	3.83
p-Hydroxy-benzoic acid	HOC_6H_5COOH	3.3×10^{-5}	4.48(19℃)
		4.8×10^{-10}	9.32(19℃)
Glycine	H_2NCH_2COOH	4.5×10^{-3}	2.35
		1.7×10^{-10}	9.78
L-Alanine	H_3CCHNH_2COOH	4.6×10^{-3}	2.34
		1.3×10^{-10}	9.87
L-Serine	$HOCH_2CHNH_2COOH$	6.5×10^{-3}	2.19
		6.2×10^{-10}	9.21
L-Threonine	$H_3CCHOHCHNH_2COOH$	8.1×10^{-3}	2.09
		7.9×10^{-10}	9.10
L-Methionine	$H_3CSC_3H_5NH_2COOH$	7.4×10^{-3}	2.13
		5.4×10^{-10}	9.27
L-Glutamic acid	$C_3H_5NH_2(COOH)_2$	7.4×10^{-3}	2.13
		4.9×10^{-5}	4.31
		2.1×10^{-10}	9.67
2,4,6-Trinitrophenol	$C_6H_2OH(NO_2)_3$	0.38	0.42

Appendix 4 Dissociation Constants for Bases (25℃)

Name	Formula	K_b	pK_b
Ammonia	$NH_3 \cdot H_2O$	1.8×10^{-5}	4.75
Hydroxylamine	NH_2OH	9.1×10^{-9}	8.04
Calcium(Ⅱ)ion	Ca^{2+}	4.0×10^{-2}	1.4
Aluminum(Ⅲ)ion	Al^{3+}	1.0×10^{-9}	9.0
Barium(Ⅱ)ion	Ba^{2+}	0.25	0.6
Sodium ion	Na^+	6.30	-0.8
Magnesium(Ⅱ)ion	Mg^{2+}	2.5×10^{-3}	2.60
Methylamine	CH_3NH_2	4.2×10^{-4}	3.38
Butylamine	$CH_3(CH_2)_3NH_2$	4.0×10^{-4}	3.40
Diethylamine	$(C_2H_5)_2NH$	1.3×10^{-3}	2.89
Dimethylamine	$(CH_3)_2NH$	1.2×10^{-4}	3.93
Ethylamine	$C_2H_5NH_2$	5.6×10^{-4}	3.25
1,2-Ethanediamine	$H_2NCH_2CH_2NH_2$	8.3×10^{-5}	4.08
		7.2×10^{-8}	7.14
Triethylamine	$(C_2H_5)_3N$	6.3×10^{-4}	3.20
Hexamethylenetetramine	$(CH_2)_6N_4$	1.4×10^{-9}	8.85
Ethanolamine	$HOCH_2CH_2NH_2$	3.2×10^{-5}	4.50
Aniline	$C_6H_5NH_2$	7.4×10^{-10}	9.13
p-Benzidine	$(C_6H_4NH_2)_2$	4.5×10^{-10}	9.35
		2.7×10^{-11}	10.57
1-Naphthylamine	$C_{10}H_9N$	8.3×10^{-11}	10.08
2-Naphthylamine	$C_{10}H_9N$	1.4×10^{-10}	9.84
p-Anisidine	$CH_3OC_6H_4NH_2$	22×10^{-10}	9.65
Pyridine	C_5H_5N	1.7×10^{-9}	8.77

Appendix 5 Standard Electrode Potential (18–25℃)

Half-reaction	$E^{\ominus}$(V)
$F_2(g)+2H^++2e = 2HF$	3.053
$O_3+2H^++2e = O_2+H_2O$	2.076
$S_2O_8^{2-}+2e = 2SO_4^{2-}$	2.01
$H_2O_2+2H^++2e = 2H_2O$	1.776
$MnO_4^-+4H^++3e = MnO_2(s)+2H_2O$	1.679
$2e+4H^++PbO_2(s)+SO_4^{2-} = PbSO_4(s)+2H_2O$	1.6913
$HClO_2+2H^++2e = HClO+H_2O$	1.645
$2HClO+2H^++2e = Cl_2+2H_2O$	1.611
$Ce^{4+}+e = Ce^{3+}$	1.72
$H_5IO_6+H^++2e = IO_3^-+3H_2O$	1.601
$2HBrO+2H^++2e = Br_2+2H_2O$	1.574
$2BrO_3^-+12H^++10e = Br_2+6H_2O$	1.482
$MnO_4^-+8H^++5e = Mn^{2+}+4H_2O$	1.507
Au(Ⅲ)+3e = Au	1.498
$HClO+H^++2e = Cl^-+H_2O$	1.482
$2ClO_3^-+12H^++10e = Cl_2+6H_2O$	1.47
$PbO_2(s)+4H^++2e = Pb^{2+}+2H_2O$	1.455
$2HIO+2H^++2e = I_2+2H_2O$	1.39
$ClO_3^-+6H^++6e = Cl^-+3H_2O$	1.451
$BrO_3^-+6H^++6e = Br^-+3H_2O$	1.423
Au(Ⅲ)+2e = Au(I)	1.401
$Cl_2(g)+2e = 2Cl^-$	1.35827
$2ClO_4^-+16H^++14e = Cl_2+8H_2O$	1.39
$Cr_2O_7^{2-}+14H^++6e = 2Cr^{3+}+7H_2O$	1.36
$MnO_2(s)+4H^++2e = Mn^{2+}+2H_2O$	1.224
$O_2(g)+4H^++4e = 2H_2O$	1.229

continued

Half-reaction	$E^{\ominus}$(V)
$2IO_3^- + 12H^+ + 10e = I_2 + 3H_2O$	1.195
$ClO_4^- + 2H^+ + 2e = ClO_3^- + H_2O$	1.189
$Br_2 + 2e = 2Br^-$	1.0873
$NO_2 + H^+ + e = HNO_2$	1.065
$HNO_2 + H^+ + e = NO(g) + H_2O$	0.983
$VO_2^+ + 2H^+ + e = VO^{2+} + H_2O$	0.991
$HIO + H^+ + 2e = I^- + H_2O$	0.987
$NO_3^- + 3H^+ + 2e = HNO_2 + H_2O$	0.934
$ClO^- + H_2O + 2e = Cl^- + 2OH^-$	0.81
$H_2O_2 + 2e = 2OH^-$	0.878
$Cu^{2+} + I^- + e = CuI(s)$	0.86
$Hg^{2+} + 2e = Hg$	0.851
$NO_3^- + 2H^+ + e = NO_2 + H_2O$	0.8
$Ag^+ + e = Ag$	0.7996
$Hg_2^{2+} + 2e = 2Hg$	0.7973
$Fe^{3+} + e = Fe^{2+}$	0.771
$BrO^- + H_2O + 2e = Br^- + 2OH^-$	0.761
$O_2(g) + 2H^+ + 2e = H_2O_2$	0.695
$AsO_2^- + 2H_2O + 3e = As + 4OH^-$	0.68
$2HgCl_2 + 2e = Hg_2Cl_2(s) + 2Cl^-$	0.63
$Hg_2SO_4(s) + 2e = 2Hg + SO_4^{2-}$	0.6125
$MnO_4^- + 2H_2O + 3e = MnO_2(s) + 4OH^-$	0.595
$MnO_4^- + e = MnO_4^{2-}$	0.558
$H_3AsO_4(s) + 2H^+ + 2e = HAsO_2 + 2H_2O$	0.56
$I_3^- + 2e = 3I^-$	0.536
$I_2(s) + 2e = 2I^-$	0.5355
Mo(Ⅵ)+e = Mo(V)	0.53
$Cu^+ + e = Cu$	0.521
$4SO_2 + 4H^+ + 6e = S_4O_6^{2-} + 2H_2O$	0.51
$HgCl_4^{2-} + 2e = Hg + 4Cl^-$	0.48
$2SO_2 + 2H^+ + 4e = S_2O_3^{2-} + H_2O$	0.4

continued

Half-reaction	$E^{\ominus}$ (V)
$Fe(CN)_6^{3-} + e = Fe(CN)_6^{4-}$	0.358
$Cu^{2+} + 2e = Cu$	0.337
$VO^{2+} + 2H^+ + e = V^{3+} + H_2O$	0.337
$BiO^+ + 2H^+ + 3e = Bi + H_2O$	0.32
$Hg_2Cl_2(s) + 2e = 2Hg + 2Cl^-$	0.26808
$HAsO_2 + 3H^+ + 3e = As + 2H_2O$	0.248
$AgCl(s) + e = Ag + Cl^-$	0.22233
$SbO^+ + 2H^+ + 3e = Sb + H_2O$	0.212
$SO_4^{2-} + 4H^+ + 2e = SO_2 + 2H_2O$	0.172
$Cu^{2+} + e = Cu^+$	0.153
$Sn^{4+} + 2e = Sn^{2+}$	0.151
$S + 2H^+ + 2e = H_2S(g)$	0.142
$Hg_2Br_2 + 2e = 2Hg + 2Br^-$	0.13923
$TiO^{2+} + 2H^+ + e = Ti^{3+} + H_2O$	0.1
$S_4O_6^{2-} + 2e = 2S_2O_3^{2-}$	0.08
$AgBr(s) + e = Ag + Br^-$	0.07133
$2H^+ + 2e = H_2$	0
$O_2 + H_2O + 2e = HO_2^- + OH^-$	−0.076
$Pb^{2+} + 2e = Pb$	−0.1262
$Sn^{2+} + 2e^- = Sn$	−0.1375
$AgI(s) + e = Ag + I^-$	−0.15224
$Ni^{2+} + 2e = Ni$	−0.257
$H_3PO_4 + 2H^+ + 2e = H_3PO_3 + H_2O$	−0.276
$Co^{2+} + 2e = Co$	−0.28
$Tl^+ + e = Tl$	−0.336
$In^{3+} + 3e = In$	−0.3382
$PbSO_4(s) + 2e = Pb + SO_4^{2-}$	−0.3505
$As + 3H^+ + 3e = AsH_3$	−0.608
$Se + 2H^+ + 2e = H_2Se$	−0.399
$Cd^{2+} + 2e = Cd$	−0.403
$Cr^{3+} + e = Cr^{2+}$	−0.407

continued

Half-reaction	$E^{\ominus}$(V)
$Fe^{2+} + 2e = Fe$	−0.447
$S + 2e = S^{2-}$	−0.47627
$2CO_2 + 2H^+ + 2e = H_2C_2O_4$	−0.49
$H_3PO_3 + 2H^+ + 2e = H_3PO_2 + H_2O$	−0.499
$Sb + 3H^+ + 3e = SbH_3$	−0.51
$HPbO_2^- + H_2O + 2e = Pb + 3OH^-$	−0.537
$Ga^{3+} + 3e = Ga$	−0.549
$TeO_3^{2-} + 3H_2O + 4e = Te + 6OH^-$	−0.57
$2SO_3^{2-} + 3H_2O + 4e = S_2O_3^{2-} + 6OH^-$	−0.571
$SO_3^{2-} + 3H_2O + 4e = S + 6OH^-$	−0.66
$Ag_2S(s) + 2e = 2Ag + S^{2-}$	−0.691
$AsO_4^{3-} + 2H_2O + 2e = AsO_2^- + 4OH^-$	−0.68
$Zn^{2+} + 2e = Zn$	−0.7628
$2H_2O + 2e = H_2 + 2OH^-$	−0.8277
$Cr^{2+} + 2e = Cr$	−0.913
$HSnO_2^- + H_2O + 2e = Sn + 3OH^-$	−0.909
$Se + 2e = Se^{2-}$	−0.67
$CNO^- + H_2O + 2e = CN^- + 2OH^-$	−0.97
$Mn^{2+} + 2e = Mn$	-1.185
$Al^{3+} + 3e = Al$	-1.676
$Mg^{2+} + 2e = Mg$	-2.372
$Na^+ + e = Na$	-2.71
$Ca^{2+} + 2e = Ca$	-2.868
$Sr^{2+} + 2e = Sr$	-2.899
$Ba^{2+} + 2e = Ba$	-2.912
$K^+ + e = K$	-2.931
$Li^+ + e = Li$	-3.0401

Appendix 6 Conditional Electrode Potential

Half-reaction	$E^{\ominus\prime}$(V)	Media
Ag (Ⅱ)+e ══ Ag (Ⅰ)	1.927	4mol/L HNO_3
Ce (Ⅳ)+e ══ Ce (Ⅲ)	1.74	1mol/L $HClO_4$
	1.44	0.5mol/L H_2SO_4
	1.28	1mol/L HCl
Co^{3+} + e ══ Co^{2+}	1.84	3mol/L HNO_3
Cr (Ⅲ) + e ══ Cr (Ⅱ)	-0.40	5mol/L HCl
$Cr_2O_7^{2-}$ + 14H^+ + 6e ══ 2Cr^{3+}+7H_2O	1.08	3mol/L HCl
	1.15	4mol/L H_2SO_4
	1.025	1mol/L $HClO_4$
CrO_4^{2-} + 2H_2O + 3e ══ CrO_2^- + 4OH^-	-0.12	1mol/L NaOH
Fe (Ⅲ)+e ══ Fe (Ⅱ)	0.767	1mol/L $HClO_4$
	0.71	0.5mol/L HCl
	0.68	1mol/L HCl
	0.68	1mol/L H_2SO_4
	0.46	2mol/L H_3PO_4
	0.51	1mol/L HCl-0.25mol/L H_3PO_4
$Fe(EDTA)^{3+}$ +e ══ $Fe(EDTA)^{2+}$	0.12	0.1mol/L EDTA pH=4–6
$Fe(CN)_6^{3-}$ + e ══ $Fe(CN)_6^{4-}$	0.56	0.1mol/L HCl
FeO_4^{2-} + 2H_2O + 3e ══ FeO_2^- + 4OH^-	0.55	10mol/L NaOH
I_3^- + 3e ══ 3I^-	0.5446	0.5mol/L H_2SO_4
I_2 + 2e ══ 2I^-	0.6276	0.5mol/L H_2SO_4
MnO_4^- + 8H^+ + 5e ══ Mn^{2+} + 4H_2O	1.45	1mol/L $HClO_4$
$SnCl_6^{2-}$ + 2e ══ $SnCl_4^{2-}$ + 2Cl^-	0.14	1mol/L HCl
Sb(Ⅴ) + 2e ══ Sb(Ⅲ)	0.75	3.5mol/L HCl
$Sb(OH)_6^-$ + 2e ══ SbO_2^- + 2OH^- + 2H_2O	-0.428	3mol/L NaOH
SbO_2^- + 2H_2O + 3e ══ Sb + 4OH^-	-0.675	10mol/L KOH

continued

Half-reaction	$E^{\ominus\prime}$(V)	Media
Ti(Ⅳ)+e ══ Ti(Ⅲ)	−0.01	0.2mol/L H_2SO_4
	0.12	2mol/L H_2SO_4
	−0.04	1mol/L HCl
	−0.05	1mol/L H_3PO_4
Pb(Ⅱ)+2e ══ Pb	−0.32	1mol/L NaAc

Appendix 7 Solubility Product Constants (18–25℃, I=0)

Compound	K_{sp}	Compound	K_{sp}
$AgCl$	1.77×10^{-10}	As_2S_3	2.1×10^{-22}
$AgBr$	5.35×10^{-13}	Ag_2S	6×10^{-30}
AgI	8.52×10^{-17}	SnS	1×10^{-25}
Hg_2Cl_2	1.43×10^{-18}	SnS_2	2.5×10^{-27}
Hg_2I_2	5.2×10^{-29}	ZnS	2×10^{-22}
$PbCl_2$	1.7×10^{-5}	Sb_2S_3	2×10^{-93}
$PbBr_2$	6.60×10^{-6}	Ag_2CO_3	8.46×10^{-12}
PbI_2	9.8×10^{-9}	$BaCO_3$	2.58×10^{-9}
PbF_2	3.3×10^{-8}	$CaCO_3$	3.36×10^{-9}
CaF_2	3.45×10^{-11}	$CoCO_3$	1.4×10^{-13}
BaF_2	1.84×10^{-7}	$CuCO_3$	1.4×10^{-10}
MgF_2	5.16×10^{-11}	$FeCO_3$	3.2×10^{-11}
SrF_2	4.33×10^{-9}	$MnCO_3$	2.24×10^{-11}
CuI	1.27×10^{-12}	$MgCO_3$	6.82×10^{-6}
$CuBr$	6.27×10^{-9}	$MgCO_3\cdot3H_2O$	2.38×10^{-6}
$CuCl$	1.72×10^{-7}	$MgCO_3\cdot5H_2O$	3.79×10^{-6}
BiI_3	7.71×10^{-19}	$NiCO_3$	1.42×10^{-7}
$CuOH$	1×10^{-14}	Hg_2CO_3	3.6×10^{-17}
$Al(OH)_3$	1.3×10^{-33}	$PbCO_3$	7.40×10^{-14}
$Cr(OH)_3$	6×10^{-31}	$CdCO_3$	1.0×10^{-12}
$Mg(OH)_2$	5.6×10^{-10}	$SrCO_3$	5.60×10^{-10}
$Hg(OH)_2$	3.0×10^{-25}	$ZnCO_3$	1.46×10^{-10}
$Hg_2(OH)_2$	2×10^{-24}	Ag_2CrO_4	1.12×10^{-12}
$Fe(OH)_3$	2.79×10^{-39}	$BaCrO_4$	1.17×10^{-10}
$Fe(OH)_2$	4.87×10^{-17}	$PbCrO_4$	2.8×10^{-13}

continued

Compound	K_{sp}	Compound	K_{sp}
$Cu(OH)_2$	2.2×10^{-20}	$SrCrO_4$	2.2×10^{-5}
$CuOH$	1×10^{-14}	$Ag_2Cr_2O_7$	2.0×10^{-12}
$Ni(OH)_2$	5.48×10^{-16}	$CuCN$	3.47×10^{-20}
$Mn(OH)_2$	2.1×10^{-13}	$CuSCN$	1.77×10^{-13}
$Mn(OH)_3$	1.0×10^{-36}	$AgSCN$	1.03×10^{-12}
$AgOH$	2.0×10^{-8}	$Hg_2(SCN)_2$	3.1×10^{-20}
$Pb(OH)_2$	1.4×10^{-20}	Ag_2SO_4	1.20×10^{-5}
$Co(OH)_2$	2×10^{-15}	$BaSO_4$	1.08×10^{-10}
$Co(OH)_3$	2×10^{-44}	$CaSO_4$	4.93×10^{-5}
$Pb(OH)_4$	3×10^{-66}	Hg_2SO_4	6.5×10^{-7}
$Pb(OH)_2$	1.2×10^{-15}	$PbSO_4$	2.53×10^{-8}
$Sn(OH)_2$	5.45×10^{-27}	$SrSO_4$	3.44×10^{-7}
$Sn(OH)_4$	1×10^{-56}	$Ag_2C_2O_4$	5.40×10^{-12}
$Bi(OH)_3$	4×10^{-31}	$BaC_2O_4\cdot H_2O$	2.3×10^{-8}
$Ca(OH)_2$	5.02×10^{-6}	$CaC_2O_4\cdot H_2O$	2.30×10^{-9}
$Cd(OH)_2$	7.2×10^{-14}	$CdC_2O_4\cdot 3H_2O$	1.42×10^{-8}
$Zn(OH)_2$	3×10^{-17}	$Ca_3(PO_4)_2$	2.07×10^{-29}
$Ti(OH)_3$	1×10^{-40}	$Co_3(PO_4)_2$	2.05×10^{-35}
Bi_2S_3	1×10^{-97}	$Cu_3(PO_4)_2$	1.40×10^{-37}
PbS	8×10^{-28}	$MgNH_4PO_4$	2×10^{-13}
Cu_2S	2×10^{-48}	$Mg_3(PO_4)_2$	1.04×10^{-24}
MnS 无定形	2×10^{-10}	$BiPO_4$	1.3×10^{-23}
MnS 晶形	2×10^{-13}	$Ni_3(PO_4)_2$	4.74×10^{-32}
α-NiS	3×10^{-19}	$FePO_4\cdot 2H_2O$	9.91×10^{-16}
β-NiS	1×10^{-24}	$Pb_3(PO_4)_2$	8.0×10^{-43}
γ-NiS	2×10^{-26}	Ag_3PO_4	8.89×10^{-17}
FeS	6×10^{-18}	$AlPO_4$	9.84×10^{-21}
Hg_2S	1×10^{-47}	$Zn_3(PO_4)_2$	9.1×10^{-33}
HgS 红色	4×10^{-33}	$Sr_3(PO_4)_2$	4.1×10^{-28}
HgS 黑色	2×10^{-32}	$Cu(IO_3)_2\cdot H_2O$	6.94×10^{-8}

continued

Compound	K_{sp}	Compound	K_{sp}
α-CoS	4×10^{-21}	$K_2[PtCl_6]$	7.48×10^{-6}
β-CoS	2×10^{-25}	BiOCl	1.8×10^{-31}
CuS	6×10^{-16}	$Zn_2[Fe(CN)_6]$	4.1×10^{-16}
Cu_2S	2×10^{-48}		

参考文献

[1] Mendham J, Denney R C, Barnes J D, Thomas M J K. Vogel's Quantitative Chemical Analysis [M]. 6th ed. Prentice Hall, 2000.

[2] Skoog D A, West D M, Holler F J, Crouch S R. Fundamentals of Analytical Chemistry [M]. 9th ed. Brooks/Cole, Cengage Learning, 2013.

[3] Harvey D. Modern Analytical Chemistry [M]. McGraw-Hill Higher Education, 2000.

[4] Christian G D, Dasgupta P K, Schug K A. Analytical Chemistry [M]. 7th ed. Wiley, 2013.

[5] Harris D C. Quantitative Chemical Analysis [M]. 9th ed. Freeman W H, 2015.

[6] David H. Modern Analytical Chemistry [M]. McGraw-Hill Companies, 2000.

[7] Hage D S, Carr J D, Analytical Chemistry and Quantitative Analysis [M]. Pearson, 2010.

[8] Ding B J. Analytical Chemistry [M]. 2nd ed. Dalian University of Technology Press，2017.

[9] 武汉大学 . 分析化学 [M]. 5 版 . 北京：高等教育出版社，2006.

[10] 彭崇慧，冯建章，张锡瑜 . 分析化学：定量化学分析简明教程 [M]. 3 版 . 北京：北京大学出版社，2009.

[11] 胡育筑，孙毓庆，分析化学 [M]. 3 版 . 北京：科学出版社，2011.

[12] 华中师范大学，东北师范大学，陕西师范大学，等 . 分析化学 [M]. 4 版 . 北京：高等教育出版社，2011.

[13] 张梅，池玉梅 . 分析化学 [M]. 2 版 . 北京：中国医药科技出版社，2018.

[14] 柴逸峰，邸欣 . 分析化学 [M]. 8 版 . 北京：人民卫生出版社，2016.

[15] 张凌 . 分析化学（上)[M]. 北京：中国中医药出版社，2016.

[16] 潘祖亭，黄朝表 . 分析化学 [M]. 武汉：华中科技大学出版社，2011.